The Housing Condition of Rural Migrant in Bejing Fengtai District

刘文韬 著

黄河水利出版社
·郑州·

图书在版编目(CIP)数据

北京丰台区农村流动人口的房屋质量调查 = The Housing Condition of Rural Migrant in Beijing Fengtai District/刘文韬著. —郑州:黄河水利出版社,2018.10
ISBN 978-7-5509-2156-6

Ⅰ.①北… Ⅱ.①刘… Ⅲ.①农村住宅-影响-农民-生活状况-调查研究-丰台区 Ⅳ.①TU241.4 ②D422.7

中国版本图书馆 CIP 数据核字(2018)第 228285 号

出 版 社:黄河水利出版社	网址:www.yrcp.com
地址:河南省郑州市顺河路黄委会综合楼 14 层	邮政编码:450003

发行单位:黄河水利出版社
发行部电话:0371-66026940、66020550、66028024、66022620(传真)
E-mail:hhslcbs@126.com
承印单位:河南新华印刷集团有限公司
开本:787 mm×1 092 mm 1/16
印张:9.25
字数:350 千字　　　　　　　　　　　　　印数:1—1 000
版次:2018 年 10 月第 1 版　　　　　　　印次:2018 年 10 月第 1 次印刷

定价:55.00 元

List of Abbreviations

BCD	:	Beijing Construction Department
BMRS	:	Beijing Migrant Registration System
BFPCO	:	Beijing Foreign Population Census Office
BG	:	Beijing Government
BHM	:	Beijing Housing Market
BMBS	:	Beijing Municipal Bureau of Statistics
BTG	:	Beijing Travel Guide
CCG	:	China Central Government
GM	:	Google Map
ICF	:	ICF International analysis of AHS data
PHCE	:	Population and Housing Census of Ethiopia
RUC	:	Renmin University of China
SCD	:	Shanghai Construction Department
UN	:	United Nations
UNHSP	:	United Nations Human Settlements Program
USDHUD	:	U. S. Department of Housing and Urban Development

Contents

List of Abbreviations
Chapter 1　Introduction (1)
  1.1　Introduction (1)
  1.2　Housing Characteristics and Rural Migrants: An Overview of the Chinese Case (3)
  1.3　The Problem Statement (4)
  1.4　Significance of the Study: A Neglected Research Area (8)
  1.5　The Scope of the Study (9)
  1.6　Thesis Structure (10)
Chapter 2　Theoretical Framework and Literature Review (11)
  2.1　Introduction (11)
  2.2　Housing Theories (12)
  2.3　Theoretical Framework (13)
  2.4　Conceptual Framework (14)
  2.5　Reviews of Literature (15)
Chapter 3　Housing System: the Chinese Perspective (44)
  3.1　Introduction (44)
  3.2　Chinese Economic Performance (44)
  3.3　The History of the Housing System Development in China (46)
  3.4　The Housing System in Beijing (60)
  3.5　Summary (64)
Chapter 4　Research Objectives and Methodology (66)
  4.1　Introduction (66)
  4.2　Research Approach/Process (66)
  4.3　Research Objectives and Research Questions (68)
  4.4　Definition (69)
  4.5　Data and Information Collection (72)
  4.6　Data and Information Analysis (76)
  4.7　Research Constraints (77)
  4.8　Questionnaire Design (78)

| 4.9 | Summary | (79) |

Chapter 5 Analysis and Results ············· (80)
5.1	Introduction	(80)
5.2	Research Area	(81)
5.3	Rural Migrant	(85)
5.4	Housing Characteristics	(96)
5.5	Influence on Living Condition	(106)
5.6	Conclusions	(113)

Chapter 6 Implication of Findings ············· (116)
6.1	Introduction	(116)
6.2	Implications on the Housing Theories	(116)
6.3	The Implications on Government Policy	(119)
6.4	The Implications on the Rural Migrants	(122)

Chapter 7 Overall Conclusion ············· (125)
7.1	Introduction	(125)
7.2	Summary of Discussion	(125)
7.3	The Future	(127)
7.4	Limitations and Scope for Further Research	(128)

References ············· (130)

Chapter 1

Introduction

1.1 Introduction

Since the last middle - century, almost all developing countries have been urbanizing swiftly. The United Nations has announced that in the next decade, 80% of the population growth will be in the urban areas of the developing nations (UN, 2007). Furthermore, by the year 2000, 45% of this population will live in urban areas among the developing countries. In China, according to the 2000 population census, the rural migrants in China have escalated to 97 000 000. According to the 2009 population census, the number of urban and rural migrants in Beijing has increased to 7 million; while two thirds of the migrants are from rural areas. The drastic increase of rural migrants to Beijing will definitely produce more burdens for the government in solving the housing problem, so the rural migrants' living condition should well be considered (CCG, 2000; CCG, 2009a).

After the national reformation in China in 1978, an excess of rural labor forces migrated to urban areas to look for better jobs, especially in big cities like Beijing, Shanghai and Shenzhen. But after 2000, millions of rural migrants surge into the urban cities. Even in some cities, the population of migrants is more than local residents. Such an increase definitely affects the national economy while enhancing the city development, especially the housing development. In China, rapid urbanization has had a tremendous impact on urban areas in terms of population growth, housing characteristics, and housing needs, especially in the metropolitan areas. In Beijing, around two million and half rural migrants surge into this second largest city in China, every year (Bian, 1994; Wu, 2010).

Rural migrants in Beijing refer to the migrants who live in Beijing for more than one month, without Beijing's urban *hukou* (household registration system), but who carry the rural *hukou* from other provinces (Wang, 2005). Chinese government use *hukou* (local ID) to define peoples' place origin and the extent to which social amenities could be accessible. Due to the number of migrants increase at a fast rate which could be viewed as burdens to most cities, therefore, the Chinese government strictly controls the *hukou* (local ID). The Chinese government does not help the migrants too much, but contrarily to wish the migrants to go back to their hometown at the end to reduce the heavy burden of big cities like transportation, housing problems, occupation competition etc. Thus, it is difficult for migrants to get the *hukou* (local ID). Migrants can be divided into two types: urban migrants and rural migrants. If the migrants come from urban areas, they are defined as urban migrants and vice – versa (Wu, 2010). This research focuses on rural migrants in the district of Fengtai in Beijing. Since they live and work in Beijing, their living conditions should be considered in their urban livelihood.

Migration is a growing urbanization activity, which has been significantly recognized by the authorities and the society as a whole. For instance, since 1995, the "Beijing Migration Control System" has officially controlled the migrants in the sense that all migrants need to be registered under the system. Then the migrants could hold a temporary license to live in Beijing temporarily (one year).

However, the majority of the migrants in Beijing have very low skill and education, and this especially applies to rural migrants. It is difficult for them to find good jobs, where most of them do some small businesses and yet, have low salaries. As a result, they are forced to live in crowded rental housing with facilities below the standard (Blau and Ruan, 1990; Meng, 2009). According to the "Beijing Housing Policy", the Beijing government does not have any housing policy for migrants because they do not have the Beijing *hukou* (Household Registration System). The migrants cannot enjoy the "Commercial Housing", "Economically Affordable Housing", "Low – rent Housing" etc. Therefore, for the migrants with low skill, low education and low salary, the only choice for them is renting houses (Wu & Wang, 2002).

In spite of its importance, academic scholars have undertaken little systematic study of migrant housing at empirical level. Subsequent micro level studies have incorporated housing characteristics and made some linkages to the living condition. Since Beijing government policy makers and authorities have neglected the importance of migrant housing, mainly because of the absence of *hukou* (Beijing ID), these groups of people have been exposed to a lot of housing problems.

The aim of this research is to examine the housing characteristics and their effects to living condition for rural migrants in Beijing Fengtai District. As Zhou (2008) puts it, the rural migrants are forced to live in these housing with many problems because of the fact that they lack the choice. Thus, these rural migrants end up renting an apartment with low rentals in the neighborhood. Thus, in cities, like Beijing, neighborhood plays some very important roles for rural migrants.

1.2 Housing Characteristics and Rural Migrants: An Overview of the Chinese Case

In Beijing, with the increase in the economy and growth of population, a lot of social problems emerge, especially the housing problem for migrants. Due to the fact that the housing rental in Fengtai is lower than other districts in Beijing, the majority of rural migrants prefer to live in Fengtai first when they migrate to this city (Wang, 2007; Chan, 1994). Also, the researcher lives in Fengtai District and quite familiar with the situation here. This justifies why this research mainly focuses on one district in Beijing—Fengtai District, and the housing problems focus on one special group only— the rural migrants.

Until 2011, the whole population in Beijing Fengtai District was 1 360 000, while the population of migrants in Fengtai District was 489 000 (BTG, 2011). The number of rural migrants was higher by 72.19% (BCD, 2012) among the total migrants (489 000) in Beijing Fengtai District. Such a large number of rural migrants living and working in Beijing Fengtai District without Beijing *hukou* (household registration system) (Zhu, 2007; Jiang, 2006), and at the same time, there are a lot of accommodation problems for this special group, especially the problems related to the housing characteristics like overcrowding, housing privacy and housing facilities (Jiang, 2006). In Beijing Fengtai District, the rural migrants usually rent apartments in the neighborhood; these houses are usually very small and the whole family would be 'squeezed' inside; some rural migrants would rent an apartment and share with other people, where they share the kitchen, bathroom and living room, so privacy is not guaranteed. Meanwhile, some other rural migrants want to save money and would not mind renting the house without good facilities, which mean no running tap water, no Internet and even without any piped gas in the kitchen. Since they have to settle in these houses, their living conditions are definitely affected (Chen, 2003; Chan, 1996).

These rural migrants come to Beijing Fengtai District for the purpose of finding better jobs to upgrade their living conditions. However, the majority of these rural migrants consider staying here for short period of time and finally they see the prospect of leaving for good (Meng, 2009). Thus, the rural migrants end up investing little money for housing improvement and instead, would go for bringing cash to their hometowns. Thus, the rural migrants' housing in Beijing is very crowded, they lack the privacy and they live without complete facilities. In this vein, these poor housing characteristics result in a bad living condition which in turn, seriously affects life and health for the rural migrants (Chen and Gao, 1993).

The target of this research is to know the real situation of rural migrants' housing characteristics and how they affect their living conditions. Due to the high housing rental in the capital, many rural migrants choose the houses with poor facilities and choose to cohabit. Thus, many rural migrants have to settle sharing one room and their houses tend to become very crowded. Therefore, their personal privacy is automatically destroyed. Therefore, three housing characteristics are considered for rural migrants in Beijing Fengtai District: overcrowding, privacy and facility. Next, these housing characteristics are evaluated to determine their effects on rural migrants' living conditions.

Although data and information on rural migrants' housing characteristics in Beijing are still scarce, studies carried out by Jiang (2006) and Wu (2010) have shed significant light in analyzing rural migrants' housing characteristics and examining the effects on rural migrants' living condition. They have shown some important elements of the housing characteristics like housing overcrowding, housing privacy and housing facility. Also, they use a very clear method to analyze the relationship between housing characteristics and living conditions.

1.3 The Problem Statement

The rapid urbanization process in the developing countries coupled with high population growth rates has pressured city development, especially on the issues of housing. The increase in urban population is related to environmental, occupational and living adaptations in urban areas, notably in the metropolis like Beijing, Shanghai and Shenzhen. Thus, the environmental problems like air and water pollution, work - related competition together with housing shortages have emerged (Chiu, 1996). However, the central government of China its housing authorities seem to have little ability to solve housing problems promptly and quickly. The exponential population growth, together with the vast influx of migrants into urban areas, has created substantial impacts on housing development. Since housing is the basic and the most important issue for human beings, housing problems need to be settled first and foremost (Davis, 1990; Zhou, 2008).

A study on rural migrants has revealed that the proportion of these people is 73% of the total number of migrants (CCG, 2010), and that in Beijing, it constituted approximately 23% of the total population (CCG, 2009c). In Beijing Fengtai District, it constituted approximately 35.9% of the total population and 72.19% among the total migrants of 489 000 (BMRS, 2010). Undeniably, there is an important reason for scholars to investigate the rural migrants' housing characteristics and the effects on their living condition. Housing characteristics signify the largest proportion of housing problems

Chapter 1 — Introduction

for rural migrants (Wu & Wang, 2002). Some researchers have examined the rural migrants' housing characteristics in Beijing, but they have never done any research on how the housing characteristics influence rural migrants' living condition.

It is important to recognize that a major problem for these rural migrants in Beijing as identified by Jian and Ye (2003), Jiang (2006) rests in the fact that their housing in Beijing is very crowded, lack privacy and have incomplete facilities. Jiang (2006) establishes that the rural migrants in Beijing live in very crowded houses and their privacy have been affected due to this poor living condition. In addition, the majority of the rural migrants' housing are deprived of basic housing facilities. Similarly, as stated by Jiang (2006) and Pang (2003) housing characteristics as aforementioned are very common. According to their research in 2000, they state that, in Beijing, nearly half of the rural migrants' housing are without kitchen and bathroom, 40% of their housing are very crowded, with 4~6 people sharing one small room, and 20% of them admitted that they did not have running water (Feng, 1999).

Wang (2007) also views that the rural migrants in Beijing have very few housing choices without the *hukou*. The Beijing urban *hukou* allows residents to purchase three types of housing with reliable financial capabilities. These three types of housing are: (1) Commercial Housing; (2) Economically Affordable Housing; (3) Low – rent Housing. However, these choices are unattainable by rural migrants since a m^2 would cost around 30,000 RMB (BHM, 2012). This automatically would make anyone understand that very few rural migrants can afford their own housing in Beijing (Gaubatz, 1995).

Wu & Wang (2002) and Wang (2007) have stated that rental housing is the most reasonable choice for rural migrants to live in Beijing. They cannot obtain other types of houses because: (1) They do not have the Beijing *hukou*; (2) They cannot afford the said houses.

A study of housing characteristics in Melanesia suggests that good housing can be interpreted by "achieving privacy and avoiding overcrowding", especially in reference to the migrant housing (Sababu, 1998). Moreover, another research in Ethiopia has stated that good housing should not be crowded and supplied with good sanitation conditions (Abera & Yemane, 2002).

Overall, the nature of the relationship between the three housing characteristics and living condition suggests that some analyses of the rural migrants' housing characteristics are very timely. In the context of this study the nature of this relationship has to be verified in order to determine the significance of rural migrants' housing characteristics and the effects on their living conditions.

The problems discussed above are directly relevant to the main areas of rural migrants' housing characteristics and effects on their living condition. Moreover, good housing characteristics play a significant role in rural migrants' lives and they can affect their living condition (Lau, 1993). In relation to this, the researcher hopes that this research could perform certain functions, which are worthy of much closer examination.

1.3.1 Issues to be Examined
1.3.1.1 Rural Migrants

Consideration of socio-economic issues will entail exploring the characteristics of rural migrants in Beijing Fengtai District. It is important to examine the profile of these rural migrants to determine their origin, gender, education level and other demographic questions so as to establish the real participants of this socio-economic activity since empirical evidences from various studies have shown that migrants are more likely to engage in such activities. The demographic questions are also necessary to ascertain the reasons behind the migration since middle and poor income people tend to be involved in migration. Nevertheless, as An (2006) has noted, not all people who migrate to Beijing are poor or are in the middle income group, and instead, very few are rich people. This characteristic is closely associated with the education level of the rural migrants. Empirically, rural migrants are uneducated, where mostly they have only managed to complete less than six years of schooling, with some attending secondary schools, few people getting the diploma certificate (Ma, 1998) which results in a lack of social skills and limited employment opportunities. Tan (2003) and Wang (2005) have noted that because of rural migrants' low education level, obtaining a good job in urban areas will be very difficult for them. The majorities of them have, in fact, had no choice and continued to embark into some small businesses of their own (Leaf, 1997; Nee, 1996).

1.3.1.2 Housing Characteristics

Migration activities may have not only developed a heterogeneous relationship with the external economy but also extended this heterogeneity within the rural migrants themselves. Following the views of some researchers (Jiang, 2006; Meng, 2009) they point out that the housing characteristics play very important roles among rural migrants, and also they highly affect their lives. The lives of those rural migrants would be dependent on several factors or a combination of those factors, with good housing characteristics standing as an indicator of good life endowment.

With the economy development in Beijing, the population increase has been rapid. Meanwhile, due to the limited land and the fact that millions of migrants surge into Beijing every year, housing overcrowding has become a common issue. Also, as many rural migrants rent the low rental, poor-facility housing and share it together, their privacy is always affected. Such a decision (to rent low – cost housing) is explained by their low economy. Among thousands of housing characteristics, the overcrowding characteristic, privacy characteristic and facility characteristic would pose serious problems for migrants, especially for rural migrants (Jiang, 2006; Wu & Wang, 2002; Wang & Wang, 2002; Jian & Ye, 2003). Evidently, solving these housing characteristics' problems would improve their housing situations and contribute to their well – being.

Nonetheless, while trying to provide solutions to the housing characteristics of rural migrants, it is also crucial to explore their life profiles first (Meng, 2009). Who are the rural migrants? Where do they live? We can use some of the examples of socio – economic characteristics that are essential to gain a better understanding of the functions involved.

1.3.1.3 Living Condition

Most studies do not pay attention on rural migrants' housing characteristics and living condition. Since research has also shown that the elevated status of living condition accompanied by high rural – urban migration results in a high growth rate of urbanization, rural migrants are found to migrate to urban areas to enhance their living condition (Lim and Lee, 1990; Wang, 2005). Living condition refers to the factors that could influence human's life and work either inside or outside of their houses (Jiang, 2006). Nonetheless, the real situation for rural migrants in China is not optimistic. A majority of rural migrants in the urban areas have very poor living condition. Furthermore, some rural migrants even do not pay enough attention to the betterment of their own living condition (Wu & Wang, 2002). Thus, to upgrade the living condition in Beijing, the rural migrants should adopt a positive attitude.

1.3.1.4 Hypothesis

Since it is assumed that migration is an alternative means of livelihood for the urban and rural migrants, it is essential for researchers to examine the housing characteristics and the effects on living condition in urban areas through an empirical analysis. Rural migrants are characterized as an unskilled and uneducated group with limited capital income. According to the theoretical framework, rural migrants live in urban areas, and there are some housing issues that are prevalent such as overcrowding, privacy and facility problems. At the same time, the rural migrants' living condition in urban areas should also be considered. It is therefore, hypothesized that the housing characteristics could influence rural migrants' living condition in Beijing Fengtai District.

1.3.2 Research Questions

(1) What is migrant housing? Where are their locations? What types of housing?

(2) Who are the rural migrants? What are their profiles? What are their occupational

history?

(3) What are the housing characteristics in Beijing Fengtai District? What are their living condition?

(4) How do housing characteristics affect the rural migrants' living condition in Beijing Fengtai District?

1.4 Significance of the Study: A Neglected Research Area

Many previous studies carried out on the rural migrants have focused mostly on housing characteristics but with no mention of their relationship with living condition. Zhu (2007) acknowledges that housing characteristics are very important to rural migrants and the characteristics affect their life, yet, the relationship between housing characteristics and living condition is ignored. The U.S. Department of Housing and Urban Development (USDHUD, 2007) also mentions the importance of rural migrants' housing characteristics in urban areas, but scholars still have not connected them to the living condition. Nevertheless, a number of studies on rural migrants were undertaken by several researchers who include Jian & Ye (2003), Lin (2003), Zhu (2007), Meng (2009).

In China, most studies, although limited in numbers, are conducted on rural migrants who sometimes include housing characteristics as part of the investigation. Examples of these studies completed in the late 90's and early 2000 included Wu & Wang (2002) and Wang (2007), but these studies were done in Shanghai and Wuhan. Beijing Fengtai District was not included. Wu's work completed in 2010 was the most recent academic research on migrants' housing characteristics in China while the work of Jiang (2006) was carried out in the context of the Hong Kong areas. His completed a study on migrants in Hong Kong, ten years ago. Although limited and increasingly outdated, these studies have provided a valuable basis for this study and assisted the researcher to understand the overall situation of the migrants' housing characteristics in China.

Although the Beijing authorities such as Beijing Construction Department, Beijing Migration Household Registration and Beijing Migration Control System have collected data on migrants (urban & rural migrants), unfortunately, less attention has been given to accomplish academic research on housing. Local authorities, which are directly responsible for the migrants (urban & rural migrants), have also carried out ad-hoc studies from time to time. Nevertheless, although the government obtained a lot of primary data, but due to the lack of organization and minimized use by the respected authorities, these data had been wasted and were rendered to be without value.

As rural migration activities, especially the relationship between housing characteristics and living condition have received very little research attention from academia and the government; this study could be regarded as a major research effort to investigate the rural migrants' housing characteristics and their effects on living conditions.

1.5 The Scope of the Study

It is the aim of this study to demonstrate the significance of housing characteristics and the effects to the living condition for the rural migrants. With reference to previous studies completed by many distinguished researchers and writers, it is well understood that rural migrants' living conditions are highly affected by their housing characteristics in the urban areas. Zhao & Yue (2005) for instance, show that the floating population in Beijing, Shi Jiazhuang, Shenyang, Wuxi, Dongwan, Kunming are suffered by poor housing characteristics, also affecting their living condition. Zhang & Hou (2009) in their study of migrants in Beijing have suggested that good living condition should be accompanied with good housing characteristics and Wu (2010) shares a similar view. The Chinese living standard however, has been improving a lot in the last decade until very recent years evident through reports of economic growth and policy reformation. In Beijing, the real estate housing has been the backbone of the nation's economy. Its share towards the nation economy had expanded from 0.3% in 1990 to about 1.5% in 2010. Because the housing developing market employs the largest proportion of migrants in Beijing, it appears that the migrants' housing characteristics should not be neglected (Lin and Bian, 1991; Wu & Wang, 2002).

Yi & Hong (2008) have both concluded from their respective studies that the migrants, especially rural migrants were here mainly in the developing countries despite the societal, economical and political changes that will occur subsequently. Both writers have agreed that with the development of the economy, the migrants' housing characteristics and living condition will improve. Furthermore, when the developing countries become highly developed, the migrants, especially rural migrants will decrease until they will completely disappear. However, the situations of the migrants, especially rural migrants' housing characteristics and living conditions are still currently unclear.

It is first of all, necessary to examine the nature of the overall housing system in China. Characteristics and profiles of the rural migrants are important variables to determine the economic status of the participants; that is, whether or not they have really low income, are poor and uneducated. Next, it is necessary to examine the linkages between housing characteristics and living condition in order to establish the nature of the relationships; in particular to prove that living condition is affected by housing characteristics. This issue is crucial for rural migrants to realize that their living condition is affected by their housing characteristics, also it could inspire the government to help the migrants the best way possible.

1.6　Thesis Structure

In Chapter 1, the researcher will introduce the overall living condition for rural migrants all over the world and then we will relate this to the similar situation in China. The research area background and research methods will briefly be introduced by the researcher. In Chapter 1, the researcher will put forth the research objectives, as well as the research background and purpose of this research.

In Chapter 2, the researcher will discuss some research areas related to the topic in detail all over the world. The theoretical framework will be introduced in this chapter. Some important cases related to housing characteristics and the living condition of rural migrants will be introduced, with a particular attention given to the work done in China. The analysis and conclusion about the present situation of rural migrants' living condition are also provided.

The importance of the housing system is demonstrated in Chapter 3, which provides some historical and present day perspectives in China. This chapter includes a discussion on the housing system in China in relation to rural migrants and the type of housing for them to live in.

In Chapter 4, the researcher will focus on the research methodology. The research approach, objectives and method of data collection as well as the limitations of the study are discussed in this chapter.

Chapter 5 includes the analysis and examination of the rural migrants' housing characteristics in relation to the living condition based on the findings of the survey. Analysis is concentrated on the rural migrants' housing characteristics including overcrowding, housing privacy and housing facility. The analysis also includes an examination of the effects of housing characteristics on rural migrants' living condition.

In Chapter 6, the researcher will discuss the implications of the findings. This is the basis for re-examining the main objectives of the study.

In Chapter 7, the researcher will conclude the study by examining the relationship of the findings to the hypothesis. Also, it addresses some other possible areas for further research.

Chapter 2
Theoretical Framework and Literature Review

2.1 Introduction

The developing nations have experienced various changes in their housing development over the years. Regardless of the development or under-development in some countries, the housing characteristics have been associated directly with people's living conditions (Jiang, 2006). The scope of the theoretical analysis must therefore encompass the relationship pertinent to the rural migrants' housing characteristics and living condition. Two major approaches have dominated the leading theories of housing development in the developing and developed countries namely the living and housing theory (Havel, 1957) and the housing development theory (Hao, 2009). These theories, which emphasize the relationships between living condition and housing, have direct implications for both, namely the housing characteristics and living condition.

Even though this research is concentrated on studying rural migrants' housing characteristics and living condition, it is necessary to emphasise that this research is part of the housing problems and in many instances, it comprises a large proportion of the housing issues that prevail (Cagamas, 1997; Wang, 1990). The concept of rural migrants in China was first introduced in 1987 (CCG, 1987), before researchers started to embark on their researches about this group. The majority of the researches are related to rural migrants' housing characteristics and living condition. Thus, many discussions on these subject matters have been produced. Among these researchers, Havel (1957) and Hao (2009) have established the housing theories that provide the foundations for future works. Realizing that a singular approach to the subject would automatically nullify another theory, the integrated approach has been brought into existence (Nattrass, 1997). These theoretical analyses and approaches have provided the conceptual framework for a lot of studies on migrants' housing.

2.2 Housing Theories

2.2.1 Living and Housing Theory

Since this theory was first developed in 1957 in France, it then became the backbone of the housing development policy for many developing and developed nations. Considering that this theory was developed on the relationship of living and housing development, it could be used in all the developing and developed nations. It is a housing theory in which the right mixture of housing development and living condition will accelerate the growth of the housing development all over the world.

The living and housing theory by Havel (1957) dictates that: Everyone should enjoy good living condition which is not limited by several housing characteristics (housing size, housing privacy, housing facility etc.), economy and law.

The relationship between living and housing is to be determined through improvement in the living condition and an increase in housing development. It is then assumed that the people, especially migrants will pay more attention to the housing development if they want to improve their living condition. Therefore, this theory represents the fact that migrants' living condition is highly related to their housing characteristics (e.g. Overcrowding, privacy and facility) in urban areas. Also, their living condition is related to their economy and law as well. Good economy could get good housing characteristics; also, good housing law could help the migrants to pursue the good housing characteristics. Due to economy and law could affect the housing characteristics; therefore, the situation of housing characteristics could affect the living condition.

However, this living and housing theory was developed during the period when Havel had given more attention to the study of urban housing. This is able to explain why this theory is more relevant in urban areas than in rural areas. Despite this theory's drawback, many researchers of housing have firmly placed this approach as a basis of their studies.

2.2.2 Housing Development Theory

Everyone, particularly the urban and rural migrants should equally enjoy the same housing system (e.g. types of housing, housing subsidy etc.) in urban areas (Hao, 2009). The housing development theory mentions that the law could separate the people into local and migrants by *hukou*, therefore, it will hamper the rural migrants from enjoying the housing system in the urban areas.

This housing development theory also refer to the international and pointed to the other countries like US, UK, Germany and France etc that every person should have the equal chance to enjoy the housing policy. It also mentions that no matter local or migrants, white or color, rich or poor, the housing system should not be bias and everyone could enjoy the same housing subsidy, housing choice and housing beneficial. This housing development theory strongly supports the humans who have the equal rights to enjoy the housing and benefit from it (Hao, 2009). In the world-view, this housing theory could help the migrants and other low or middle level people to get their own rights of the housing policy and make this world more harmonious.

This theory, which is based on the housing system development approach termed as "Chinese housing development plan" (CCG, 2006), integrates the modern housing theory and the statistical analysis to depict the housing system development that a developing nation must undergo in order to create a good housing system. Scholars who are associated with this theory are more radical in the political and housing system development in urban areas. Basically, this housing theory is applied to the housing system development in China and some relationships are formed between China's condition and the international settings. The target of this housing theory is to execute a better planning for the Chinese housing system and theoretically, for the transit from a developing to a developed stage can be achieved at a faster rate.

Due to this housing development plan mentioned that everyone, particularly the urban and rural migrants should equally enjoy the same housing system in urban areas. It against the housing law in China that people could only enjoy the housing in their hometown, but for the social stable and *hukou* (local ID) limitation, people are not encourage to migration and cannot enjoy the housing in other places in China. Therefore, this housing development theory could strongly support the migrants to pursue their housing rights in urban area and give more confidence to housing researchers to carry out the research.

2.3 Theoretical Framework

Based on the "Living and Housing Theory" (Havel, 1957) and "Housing Development Theory" (Hao, 2009), the housing characteristics, economy and law could affect human's living condition. The "Living and Housing Theory" mentioned that the economy, housing characteristics and law could affect people's living condition. Also, good economy and law also could improve the housing characteristics, therefore, it will affect people's living condition. The living condition refers to the factors that could influence personal and working aspects of human either inside or outside their homes (Jiang, 2006). Good living condition should have good housing characteristics, also it must be supported by robust economy and law. For the "Housing Development Theory", it mainly focuses on the housing law and this theory against the housing law could restrict the migrants to pursue the good housing in urban area. Law, translated into the legitimization of *hukou* in China enables the division of residents into local and migrants, therefore, the rural migrants could not enjoy the housing in urban areas. Finally, it could affect the rural migrants' housing characteristics and living condition. Therefore, the "Housing Development Theory" could support the migrants to pursue good housing in urban area and encourage improving the housing system in China and International to benefit everyone to enjoy the same housing right. Therefore, the migrants' living condition could be improved. Also, this theory could give more confidence to the housing researchers to carry out the research (refer to Fig. 2. 1).

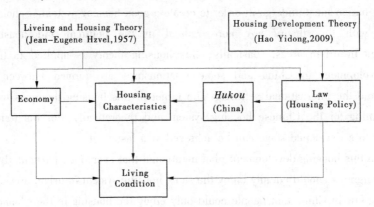

Fig. 2. 1 **Theoretical framework**

In this research, three housing characteristics in Beijing influence the rural migrants' living condition to a great degree (Wu, 2010). Housing Overcrowding, housing privacy and housing facility are three factors could determine whether rural migrants' living condition is good or otherwise. As mentioned by Jeanne and Bonnie (2001), good living condition is highly affected by housing characteristics especially housing overcrowding, housing privacy and housing facility. Without good housing characteristics, human's living condition (living, working, sleeping etc.) will be disturbed. In Beijing, good living condition must have these three housing characteristics. Thus, to determine the influence of housing characteristics on rural migrants' living condition, the study has placed an emphasis on these three housing characteristics (Wu & Zhang, 2002).

2.4 Conceptual Framework

Based on the theoretical framework, the economy and law could affect the housing characteristics and then influence human's living condition. Due to the rural migrants' low economic foundation and their inability to enjoy the housing system in Beijing, it will deteriorate their housing characteristics. In turn, the housing characteristics could influence their living condition in Beijing.

The concept framework of this research is to determine the influence of housing characteristics to rural migrants' living condition in Beijing Fengtai District. The researcher will focus on the rural migrants' housing characteristics (Overcrowding, housing privacy and housing facility) and some examination will be carried out. Then, the researcher will examine the housing characteristics which are believed to be able to influence rural migrants' living condition (refer to Fig. 2.2).

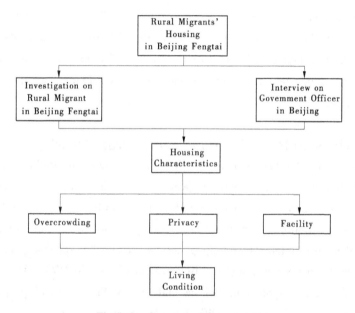

Fig. 2.2 Conceptual framework

2.5 Reviews of Literature

Recently, China's population's migration activities have become a prominent social phenomenon. Beijing, as the capital and fast-developed metropolis, attracts a large number of migrants, especially the rural migrants. Using data from one per thousand floating population investigation conducted in 2006, until the end of 2006, the total migrants in Beijing were 3.577 million, an increase of 1.012 million compared to the population recorded in 2000 (see Table 2.1). The average annual increase was 202 thousand; the average annual growth was 6.9% (BMBS, 2006).

Table 2.1 The migrants' population number in recent years in Beijing

Year	Population (million)
2000	1.012
2006	3.577
2010	6.790
2011	7.258
2012	8.630

Source: *Beijing Municipal Bureau of Statistics (BMBS)*, 2012.

The total population in Beijing Fengtai District is 1.36 million and the area is 306 km^2, while the number of rural migrants is more than 72.19% of the total migrants (or 489 000) in Beijing Fengtai District (BCD, 2012). The reviews cover the structure, distribution, economic status, housing characteristics and living condition of the rural migrants in Beijing.

2.5.1 Reviews on Rural Migrants and Migrant Housing

Therefore, with the fast increase of the rural migrants in Beijing, concentration should be dedicated to this group of people. According to the "Beijing Migrant Registration System", rural migrants refer to the migrants who live in Beijing for more than one month, without Beijing's urban *hukou*, but they happen to have the rural *hukou* from other areas. The rural migrants in Beijing need go to the "Beijing Migrant Registration System" to register every year, and the Registration fee is 170 RMB. This policy could easily define the rural migrants in Beijing and control this group of people (Semi-structured Interview For Government Officer by the researcher, 2013).

Due to the fact that Beijing is the capital city of China, the migrants tend to have some special feelings (like) towards Beijing. They think it is very honorable to migrate to Beijing and work there. Therefore, around 70.0% of the rural migrants in Beijing have chosen to migrate to Beijing only and without even thinking about leaving for other cities (Harris, 1991; Logan, 1996; Wang, 1992; Wang, 2007). These rural migrants choose Beijing as the destination, and it mirrors their high level of loyalty towards Beijing.

Even though migration is a complex process, millions of rural migrants still choose to embark into migration. Two thirds of the rural migrants migrate to Beijing due to their own willingness and the main reason for their migration is to earn money and bring back to the money to their hometown; the proportion was 71.0% according to the floating population survey in Beijing conducted in 2006. Therefore, due to the good economy Beijing offers, many rural migrants were attracted to this place. Their low education level and lack of skills, together with the fact that they do not have the *hukou* definitely impede them from finding suitable jobs in Beijing.

Due to the limitation of education and skills, more than 80.0% of the rural migrants work in the service industry area or do some small businesses. These occupations in Beijing usually have low income and do not have any guarantee. Based on the previous population research, the rural migrants have been characterized by several aspects and it has somehow altered a great deal in the last ten years (Duan, 2009).

According to the previous researchers (Zhou, 1996; Duan, 2009; Wang, 1996), more female had joined the migration. From 1994 until 2006, the rural migrants' gender ratio between male and female in Beijing decreased from 173∶100 to 123∶100. This situation had seen more and more female rural migrants following their husbands to come to Beijing and stay together. As they cannot enjoy the housing system in urban areas and only manage to rent houses with poor housing characteristics, these circumstances have severely affected their living condition. Also, most of the female rural migrants do some small businesses together with their husbands, and in time, have no time to take care of their children and family. It is worth noting that their children need to be educated, and the female migrants should pay attention to their children instead of their business (Zhou, 1996; BFPCO, 1997; Duan, 2009; RUC, 2006).

With the increase of female rural migrants, more and more children are only obliged to follow their parents to these urban areas. From 1997 to 2006, the children ratio among the total rural migrants in Beijing increased from 6.7% to 14.2%. Most of the children's age was below 14 and at this point, needed good education (BFPCO, 1997). However, without Beijing's *hukou*, they simply cannot attend the school. Therefore, most of the rural migrants' children have to quit school and live their lives without any education. Also, children's health is also vital. However, due to the poor housing characteristics and living condition, their health has become seriously affected.

Until 2006, 78.0% of the rural migrants in Beijing were already married. Among the rural migrants, 75.3% of rural migrants had brought their spouses to Beijing and lived together (RUC, 2006). As mentioned by Lu (2010), in China, the rural migrants prefer to marry early and the male has the responsibility to support his family. However, with too many rural migrants bringing their whole families to Beijing, it will definitely increase the families' burden and worsen their living condition.

According to "Beijing Migrant Registration System" (BMRS, 2010), the rural migrants' education level is very low. The majority of the rural migrants' education level has completed their secondary school, and only 6.0% of the rural migrants obtain the diploma certificate. 18.9% of the rural migrants finish their education at primary school level, whereas 4.4% of the rural migrants never attend school. Wu (2010) states that the low education level will definitely discourage the rural migrants from finding good jobs in urban areas. Next, their low income will affect too their housing characteristics and living condition.

Therefore, due to the rural migrants' low education level and inability to find suitable jobs, it is very difficult to get Beijing *hukou* for this group of people. The Beijing Government only serves *hukou* for migrants if they could get high position jobs in government departments or hold positions in big companies with high income (BG, 2006). Thus, due to the *hukou* problem, the rural migrants cannot enjoy the housing and social welfare in Beijing. Finally, their living condition will be affected.

With the *hukou* problem, the majority of (68.0%) the rural migrants stay in Beijing for around three years and the average living period is three years and a half (BMRS, 2010). Only less than 10.0% of the rural migrants live in Beijing for more than seven years. Due to the Beijing government did not concern this situation, therefore, the migrants are quite difficult to get the Beijing local ID and enjoy the social benefits in Beijing. Thus, to merge these rural migrants into Beijing social welfare, the granting of the Beijing *hukou* should be considered (Zhu, 2007).

Due to the *hukou* and economic limitation, the rural migrants in Beijing do not have too many choices for housing. According to the housing system in Beijing, the Beijing local residents could enjoy three kinds of housing: commercial housing, economic affordable housing and low-rent housing (BG, 2012). Without the *hukou*, the rural migrants simply cannot enjoy these housing types. There are more than 90.0% of the rural migrants who have chosen to rent houses in the neighborhood in Beijing. The other rural migrants have chosen to live in their friends' houses or hostels provided by the company (Wu & Wang, 2002). As more than 90% of the rural migrants choose to rent houses in the neighborhood (Zhang & Hou, 2009), as compared to the local residents, rural migrants tend to have to experience more housing problems.

Overall, the rural migrants in Beijing have too many housing problems and these problems need to be addressed soon. Due to the low economy and without local ID, the rural migrants cannot enjoy the housing in urban area. Therefore, it deteriorates their housing characteristics in urban area especially the overcrowding, privacy and facility problem. Thus, their living condition will be decreased due to the bad housing characteristics. Good living condition should be in harmony with good housing characteristics. Bad housing characteristics will definitely affect the rural migrants' living condition. Therefore, it is necessary to carry out the research and examine the relationship between housing characteristics and living condition for the rural migrants.

2.5.2 Reviews on Housing Characteristics
2.5.2.1 Overcrowding Characteristic

Overcrowding is a human reception that is stimulated by social activities and represented by housing characteristics (Altman, 1975; Baum and Davis, 2006). Some other researchers define overcrowding as a relative availability of a space in terms of area and number of room in a given (Abera & Yemane, 2002; Wu, 1996). According to the National Population and Housing Census in Ethiopia in 1984, 39% of the urban population accommodates very crowded houses (PHCE, 1984). Now, the overcrowding problems have become a common issue in the urban areas especially in big cities like New York, London, Tokyo, Hong Kong, Beijing and Shanghai. Therefore, it is necessary for the researcher to carry out this research.

Overcrowding was a problem that started from the World War II and has steadily become a social problem in urban areas up until now (Freedman & Buchanan, 1972). Due to the fast increase of the population, overcrowding in urban areas is a very serious issue. Also because of urbanization, many rural migrants migrate to urban areas and this migration further aggravates the housing overcrowding problem. According to the United Nations (2007), the overcrowding problem in urban areas becomes much more severe in the last 20 years especially in the developing countries. In some big cities like Beijing and Shanghai, four or five people sharing a room are very common. Therefore, due to the overcrowding problem, it seriously disturbs people's sleep, work, personal privacy and so on (Alwash and McCarthy, 1988; Babbie, 1983; Baldassare, 1999; Xie, 1996).

The most common measure of overcrowding is to look into person – per – room or person – per – bedroom in a dwelling unit (Blake& Kellerson & Simic, 2007). Also, before starting the research, it should be noted that people define overcrowding by their common sense, including by the number of persons in a housing, housing size, the ratio of persons to floor space in square feet, location, lot size and structure type (Ambrose, 1996; Baron, 1996; Yang and Wang, 1992).

In China, overcrowding is defined as persons – per – bedroom is a standard of more than two people (Wu, 2010). Also for person – per – room standard, the most acceptable common sense is not more than two people in a room in China. And it could help the researcher to measure the overcrowding in this research.

According to the "Traditional and Alternative Definitions of Overcrowding" (Blake& Kellerson & Simic, 2007), there are four ways to measure overcrowding:

(1) Persons – Per – Room (PPR)

(2) Persons – Per – Bedroom (PPB)
(3) Unit Square Footage – Per – Person (USFPP)
(4) Persons – Per – Room (PPR) by Unit Sq Foot – Per – Person (USFPP) (see Table 2.2)
(Source: ICF International analysis of AHS data, 1985~2005)

Based on this research methodology and research limitations, the researcher has decided to use Persons – Per – Room (PPR) and Persons – Per – Bedroom (PPB) for this current research. The bedroom is typically used for sleeping, while for the definition of a room, the UNVHS (habitat, 1991) defines it as a

"Space in a housing unit or other living quarters enclosed by walls reaching from the floor to the ceiling or roof covering, or at least to the height of two meters and of a size large enough to hold a bed for an adult, that is, at least four square meters".

(*pp.* 281)

In comparing the Chinese overcrowding perspective (Wu, 2010) to the "Traditional and Alternative Definitions of Overcrowding", there are some differences between the two housing overcrowding measurements. For Persons – Per – Bedroom (PPB) measurement, it is similar to the Chinese measurement that a bedroom should not accommodate more than two persons. Meanwhile for the Persons – Per – Room (PPR), the Chinese measurement suggests that it is not more than two people, but the "Traditional and Alternative Definitions of Overcrowding" define a room as accommodating not more than one people.

However, to consider the serious overcrowding problem in big cities of China, the researcher has decided to use the measurements that are more suitable to Beijing. Persons – Per – Room (PPR) and Persons – Per – Bedroom (PPB) accommodating more than two people will be defined as overcrowding.

Huang (2003) states that rural migrants' rented houses in the urban areas of China are indeed, very crowded. He has pointed out that more than 70% of rural migrants' houses accommodate more than three people in a bedroom and the living rooms also tend to be used as a bedroom. Zhai and Zhang (2003) further point out that overcrowding usually happens in the urban areas, especially within the rural migrants' houses. In 1995, the data from Beijing and Shanghai have shown that rural migrants' average living areas in the two cities are only 7.5 and 9.0 m^2, while the local residents' living areas are three times more than those of the rural migrants (CCG, 1995).

Table 2.2 Overcrowding standards for PPR and PPB included in the UK ODPM report

Measyre	Discussion of measures and standards	% of overcrowded households, using AHS national data		% point change since 1985
		1985	2005	
Persons – Per – Room(PPR)	This measure was the one most frequently seen during our literature review. The UK ODPM report reports standards ranging form greater than 0.75 to greater than 1.50	2.82	2.41	(0.41)
	We defined overcrowding as more than one persons – per – room. The percentage of households considered overcrowded is at the right. (We also present the percentage of households overcrowded when PPR exceed 1.50, which is shown after the one persons – per – room standard)	0.82	0.63	(0.19)
Persons – Per – Bedroom (PPB)	The UK ODPM Report also included PPB as a measure of overcrowding and it reported a standard of two persons – per – bedroom. We learned form speaking with Mr. Joe Riley about Public Housinf Authorities (PHA) and overcrowding that generally PHAs try to keep to two or fewer people – per – bedroom. (There is guidance bnout who cannot, the circumstances of sharing, etc.)	3.25	2.65	(0.60)
	With the PPB measure, overcrowding occurs as. values increase (e.g. a unit with 6 people and 2 bedrooms is considered more ceowded than a similar unit with only 4 people and 2 bedrooms). We used a standard of two persons – per – bedroom			

Continue Table 2.2

Measyre	Discussion of measures and standards	% of overerowded households, using AHS national data		% point change since 1985
		1985	2005	
Unit Square Footage – Per – Person (USFPP)	Square footage is a tangible measure of crowding and is important when considering air – bome disease. The reason being that, all else held constant, human proximity is the key to disease transmission We defined an overcrowing standard of 165 square feet per person. This standard was chosen because it produced a level of overcrowding equal to the 2.4 percent of the households overcrowded for PPR when using the 2005 AHS National data	3.00	2.44	(0.56)
Persons – Per – Room(PPR) by Unit Sq Foot – Per – Person(USFPP)	This measure is a mix of PPR and USFPP. We did across – tabulation of PPR and USFPP, using our standards of more than one person and 165 square feet. We felt this was an important measure because it highlights how households considered under one measure might not be under another. This cross – tabulation can yield a more accurate picture of the populations who are overcrowded and the degree that they are overcroeded	1.10	0.90	(0.20)

Note: Negative values are shown in parentheses.
Source: ICF International analys of AHS date.
Source: Blake& Kellerson & Simic, 2007.

Chapter 2 Theoretical Framework and Literature Review

Adequate housing should be able to achieve privacy and avoid overcrowding. Some researchers tend to equate overcrowding with space, because overcrowding always refers to the high living density (Ying, 1998). Stokols (2002) refer to overcrowding as a psychological concept with a motivational base. He has also stated that

"*Overcrowding is a personal reaction, not a physical variable also it is a personal feeling and in relation with spaces*".

(*pp.* 208)

Therefore, due to the high living density in big cities, the overcrowding problem can seriously affect people's living conditions which include sleeping, working etc. Due to the overcrowding, people tend to feel unhappy, worried or even have some psychological problems. Renting private houses is a key housing choice for rural migrants (Wu & Wang, 2002; CCG, 2009b). Wu also states that the rural migrants tend to save cash and many people share houses. Thus many rural migrants live together leading to the congested housing.

In China, rural migrants usually occupy very less space compared to urban local residents. Each rural migrant only occupies one third of the average housing area compared to local residents. Stated by Ding (2005) the migrants, especially rural migrants usually live in super crowded housing in comparison to their own homes in their hometowns. Two researchers conducted their research in United States between 1980 and 1989 and they found that accommodating more than one people per room is common (McArdle & Mikelson, 1994; Ye, 1993). Krivo (1995) states that migrants live in crowded houses because they simply have no other choices.

In conclusion, the prevalence of overcrowding among rural migrants may be attributable to several factors and show the clear direction for the researcher to carry out this research. First off, the finance limitation has forced the rural migrants to migrate in cities and rent small houses, thus overcrowding cannot be avoided. Second, the rural migrants adopt some economic strategies to bring money to their hometown and save on daily expenses. For the rural migrants, overcrowding is not necessary for them (Knodel, Chayovan, and Siriboon, 1992; Myers, Baer, and Choi 1996; Pader 1994).

2.5.2.2 Housing Privacy Characteristic

For housing privacy characteristics, the function of housing (residential only or residential, work and others) could seriously affect human's personal privacy. Housing privacy is related to overcrowding. Limited personal space could definitely affect residents' privacy. If many people live in a very crowded house, the individual privacy cannot be guaranteed. Also, some people as also use their houses for business or as storage, and thus the residents' privacy will definitely be disturbed (Chan, 1997; Zhang, 1997; Valins, 2001).

Westin (1970) has defined the privacy concept and stated that the personal privacy has four types. They are solitude, intimacy, anonymity and reserve. Altman (1996) states further that the four types of privacy could affect human's satisfaction level. He has made another point that personal privacy inside housing is definitely related to overcrowding. Later, many researchers have declared and supported this argument that personal privacy inside housing can be affected by overcrowding (Smith, 1991; Bharucha and Kiyak, 1992; Leopore, 1992). Walden (1981) has carried out a housing privacy research and concluded that residents' housing privacy is affected by the crowded state.

Chan (1997) investigates the residents' housing privacy level in Hong Kong and is able to derive that Hong Kong people's housing privacy is much affected by overcrowding.

Generally, the lack of housing privacy can give living condition negative side effects. If one's housing privacy is affected, the people will suffer from emotional reaction (Epstein, 1994; Massey, 1986). Housing privacy could define a self boundary or private area for residents. If residents' housing privacy is affected, their living condition will be affected either (Insel & Lindgren, 1998).

Bateson (1998) and Epstein (1994) have done housing privacy research and suggested that housing privacy should be maintained and protected, but if residents' housing privacy is invaded, people's physiological needs will be affected. Thus, people will feel shameful and embarrassed. Finally, human's living condition could be affected.

In China, the rural migrants live in a very crowded house, so their housing privacy will definitely be compromised. Ma (1998) conducts a housing privacy research in Beijing and declares that many rural migrants tend to use their houses for business purposes. They indicate that two thirds of the rural migrants' housing privacy are seriously affected in Beijing. Meng (2009) has done a research in Shanghai, stating that more than 76.7% of rural migrants in Shanghai rent houses and share with other people. Among these rural migrants, more than 12.8% of rural migrants use their home space to do business. She also points out that because of rural migrants' poor financial state, the majority of rural migrants do not care about the housing privacy problem. In China, some big cities like Beijing, Shanghai and Guangzhou, the rural migrants just need a place to live without any consideration on other aspects of life. Therefore, the researcher could consider these factors in this research.

In conclusion, the previous researches on housing privacy illustrate the heterogeneity of the housing privacy and the variety of perceptions held by academic scholars and researchers. The similarities among these housing studies represent that rural migrants' housing privacy is highly affected by overcrowding. Therefore, so many previous studies both prove that the lack of housing privacy could affect their living condition and it show the clear path for the researcher to carry out this research.

2.5.2.3 Housing Facility Characteristic

Good living condition should have good housing characteristics (Havel, 1957). For housing facility characteristics, having a kitchen, cooking fuel, running tap water, private bathroom, bath or shower etc. will be considered to determine whether the housing facility characteristic is good or not. Without these housing facility support, human's normal life will be disturbed (Wu, 2010; Zhou and Logan, 1996). Therefore, the researcher could pay attention to these housing facility factors and make use in this research.

In the last twenty years, with the economic and technology development, people's housing facilities have improved dramatically. In the urban areas, the local residents are willing to improve their housing facilities as the city is already their hometown. Meanwhile for the migrants, especially rural migrants, the living condition is relatively worse than that of the local residents. The absence of a kitchen, a bathroom, running tap water is very common among rural migrants in the cities. In fact, the poor housing facilities among rural migrants have two reasons: (1) Poor finance among rural migrants; (2) Temporary period of stay in urban areas and the migrants' wish to send money back to hometown (Zhang & Hou, 2009).

Since from the last century, many countries like the US, the UK, Germany, and China have had to address serious housing facility problems, especially with regards to the rural migrants. With the fast urbanization and hundreds of thousands of rural migrants surging into urban areas, their housing facility characteristics should be given due attention.

In the nineteenth century, during the Industrial Revolution period, many peasants were forced to migrate to the urban areas. Confronted by so many rural migrants, many cities in the UK were not ready to provide enough and adequate housing for these people. During these periods, these rural migrants' living condition was very bad. Their housing was very crowded and the housing facility was extremely poor – without a kitchen, bathroom even electricity. Because of the facility shortages, the number of people living in a small room with the size less than 10 m^2 increased from 7.03 in 1801 to 7.72 in 1851 and even to 7.85 in 1881. This situation continued until 1891 (Stanley, 1971; Zhu, 1993).

The Industrial Revolution in Germany started from 1930 until now. Finally, Germany became the second strongest economic country with high level of urbanization. Due to the urbanization, many rural migrants had migrated to the urban areas. In 1950, the government had tried to solve the housing problems for these rural migrants and built houses for them, although with poor housing facility. These houses came with bad sanitation, layout disorder and overcrowded. Therefore, these rural migrants' mental and physical healths are very much affected by the bad housing facilities.

As a result of the urbanization process, some western countries like the US, the UK, Germany etc have built many types of housing for urban and rural migrants in the last century. According to the housing style and migrants' working type, the migrant housing could be divided into three types: migrant worker housing, migrant community and colored people community (Wang & Huang, 1999). As the government did not care about the migrants in urban areas, their housing facilities in these houses are in a deteriorating condition.

a. Housing for migrant worker

This type of housing was built in the nineteenth century and was used to cater for the migrant workers. Many countries like Germany, UK and US etc. built this type of housing In the western countries, as the governments wanted to solve the migrant workers' housing problems, they had invested a certain budget and built these poorly - conditioned housing. In that period, the migrant workers' housing was described as "overcrowding" and "nasty". The migrant workers' housing was built without a private bathroom, with more than ten families sharing one toilet outside their rooms. Also, these houses have one third of the area located underground, thus the houses were very humid and had poor air ventilation. Usually there were 18 ~ 20 people squeezed inside one house (Li, 2004). From 1780 to 1850, the migrant worker housing was located in the western countries and based on multiple factors, their health had been severely affected.

b. Migrant community

From 1860 to 1920, the urbanization process increased rapidly in the world. During these years, many migrants especially rural ones had rushed into the big cities. Thus, many countries like the US, UK, Germany and France had built the migrant community for these people. These migrant communities were very packed and housing facilities had not been maintained at all. Lack of running tap water and the fact that the sewerage system was not working had prevailed in the migrant communities. In 1911, the US government reported that "the migrant communities were not suitable to live because the government did not provide adequate housing facility maintenance". In Germany, there were 7.3 million migrants living in the migrant communities (Zhou, 2008). Overall, the migrant community provides a shelter for migrant, but also they suffer a lot, so as to live inside the migrant community.

Chapter 2　　　　　Theoretical Framework and Literature Review

c. Colored people community

In the nineteenth century, many western countries had made up a community for colored people. The most famous colored people community was in Manhattan, US. In 1930, there were 164.5 thousand colored people residing inside these communities. The houses were without any maintenance, and there were more than 11% of houses going to collapse. 23% of the houses in the community were damaged severely. Many glasses were broken and no people had the urge to change the panes. Therefore, the colored people's housing facility and living condition cannot be imagined in that period (Stanley, 1971).

However, in the last ten years, many countries have upgraded the public housing for migrants in the urban areas, and provided houses with proper housing facilities (Wang, 2014). Compared with the previous migrant housing, rural migrants' housing facilities have improved a lot in the western countries. In the US, the governments have encouraged the effort to build the public housing for migrants and low – income residents. To build the public housing, the US government decreased tax and provided free land for housing developers. Also, the government provided 10.0% to 20.0% of the housing subsidy to build and renovate the public housing. With the help of the US government, the migrants and low – income residents live in the houses with good housing facilities.

Compared with the US, the UK government has provided high housing subsidy for the developers to build the public housing especially for the migrants and low – income residents. Sometimes, the housing subsidy can reach up to 30.0% of the total housing price. The migrants and low – income residents could rent the public housing with very low housing rental. This public housing policy is also implemented in France, Germany and Spanish. In France, among one thousand people, 70 people live in public housing. In Germany and Spanish, the public housing even reaches 60.0% of the total houses in some areas.

In Singapore and Hong Kong, the number of public housing is very high. Due to the high housing price and limited land, many people need to rent the public housing with low rental. These public houses are supported by the government and through social donation. In Hong Kong, around one third of the population dwells in public houses. In Singapore, around 82.0% (3.1 million) of the population live in the public houses (Wang, 2014).

For the last ten years, these countries have successfully carried out the public housing system to help the migrants and low – income residents. These public houses provide these people with low rental and proper housing facilities. With the help of the public housing, the rural migrants could enjoy better living condition in urban areas.

Nevertheless, the public housing system did not carry out well in China. Without the government subsidy and with large population, the housing developers are not willing to build the public housing. Also, it is affected by the *hukou* limitation in the urban areas. Therefore, the migrants cannot benefit from the housing system in China. In the Asian countries, the rural migrants prefer especially among developing countries migration. In China, the largest population creates a frequent population movement. Since China has more than 970 million peasants (CCG, 2000), the majority of migrants are rural migrants. As many rural migrants migrate to urban areas and work there temporarily, the housing problem will definitely emerge.

In Shanghai, the housing facility survey in 2006 represented that rural migrants' housing facilities were very poor. More than ten rural migrants live in one house and without any housing privacy. Also, the majority of rural migrants are without private kitchen, bathroom and hot water supply. 67.0% of rural migrants are without private kitchens, 71.0% of rural migrants are without bathroom and more than 83.4% of rural migrants have had to adapt to the absence of hot water supply (see Table 2.3).

Table 2.3　Rural migrant housing facility in Shanghai in 2006　　(%)

	Kitchen	Bathroom	Hot water supply
Private	19.5	13.9	8.8
Shared	13.5	15.1	7.8
Not available	67.0	71.0	83.4
Total	100.0	100.0	100.0

Source: *Data from the rural migrant housing facility survey in Shanghai conducted in* 2006. SCD (2006). Shanghai Construction Department.

The capital city, Beijing has more than five million rural migrants and the majority of them are renting (CCG, 2009c). The housing facility problems not only affect their living conditions, but also affect their mental and physical health.

According to the 2006 rural migrant housing facility survey in Beijing, it represented that the rural migrants' housing facility was very poor. 63.1% of rural migrants shared the water supply and 4.8% of rural migrants were without supply, therefore, many rural migrants had to take the water from the outside. 36.8% of rural migrants had survived without kitchen, so they had to cook inside their rooms. 10.8% of rural migrants had been without any stove heating in the winter, so their rooms were very cool. 72.9% of rural migrants shared bathroom and 23.1% of rural migrants had survived without bathroom; more than 75.4% of rural migrants were not able to take a bath in their houses, and instead they had to take a bath in the public wash houses (see Table 2.4).

Table 2.4 **Rural migrant housing facility in Beijing in** 2006   (%)

	Kitchen	Bathroom	Water supply
Private	30.2	4.0	32.1
Shared	33.0	72.9	63.1
Not available	36.8	23.1	4.8
Total	100.0	100.0	100.0

Source: *Data from the rural migrant housing facility survey in Beijing conducted in* 2006. BCD (2006). Beijing Construction Department.

Among the developing and developed countries, the rural migrants usually tend to migrate to the big cities and live there temporarily. Therefore, the housing facility problems will gradually surface for this group. Due to the fact that the public housing system is carried out well in many countries, the rural migrants could enjoy the better houses with good facilities. Meanwhile in China, due to the lack of attention given by the government and together with large population, the public housing system is still under-developed. To improve the housing facility for rural migrants, China should do something to improve the public housing system. Thus, the rural migrants could enjoy living in better houses to improve their overall living condition.

2.5.3 Reviews on Living Condition
2.5.3.1 Introduction

This section talks about the housing living condition in the other countries, from those of the Asian region to the rest of the world. This section will also highlight the living condition in these countries previously, also how these countries improve the people's settlements and their overall living conditions. The summary will be discussed in the final part of this section.

With the economic development, living condition has improved in time and therefore, people's demand for their housing characteristics has gradually been greater than ever before. Generally, people want more than just a living space, where they require sub-rooms; people would rather have a smaller construction area and more individual rooms; and the kitchen is expected to have a door, as to not let the smoke seep into the other rooms in the house. People also tend to need a storage space in the kitchen, to have a separate place to eat; the space of the toilet would be better if a bathtub is included and so on. These requirements are very simple, where they constitute very basic human needs. However to the rural migrants in China, especially the rural migrants living in the capital of Beijing, people can hardly enjoy these. It is something that they so desire, but simply unattainable. So, the majority of people's living condition has suffered, especially in Beijing. They can hardly buy a house, and instead they have to either pay a great deal monthly to rent a small house, or share with other people in a mansion. From the previous research (Zhou, 2009), only less than 5.0% of rural migrants in Beijing own houses. With such a small percentage, it is easy to sum up at this point that the living condition in Beijing necessitates serious development in the future.

In 2008, it was found that the majority of the urban residents experienced very poor living conditions in Beijing. Over 27% of the urban populations shared their dwellings with others, and 7.4% had an average living space below 4 m^2. Compounding the situation is the poor housing quality. About 76% of the urban residents did not have their own toilets, 37% had to share their kitchens with others, and 27% admitted to not have running water(CCG, 2009a).

It is very difficult to characterize 'housing' with only one precise definition—it is an element of the community's dwelling area with its own social and spatial environments, in which exist many different forms of houses, different architectural styles and designs. These years in China, because of the population and economic growth, housing becomes the most expensive asset; usually it will cost decades or even one's whole life to earn money to afford to buy a house. It satisfies one of the major personal needs, thus this motivates people to work, do something for their lives, study etc. Nevertheless, because of the land limitation and population growth, housing price soars, especially in the capital city, Beijing. Until now, the majority of people still live in inadequate housing. Only a few people could afford the place in which they live and have a good life. While most of the people still live in a bad situation, a lot of problems are created surrounding this aspect.

For almost 30 years of exploring and researching, the researcher (Fu, 2009) has established a housing living condition standard named the "International Living Condition Standard" and used all over the world. This standard was introduced in 1976 and has become the most important yardstick in housing. The "International Living Condition Standard" includes four characteristics:

Humanity characteristics. The housing should suit people and provide them with good living condition. Also it should satisfy people's mental and physical needs, to enable them to live their lives with convenience.

Social characteristics. Good housing represents social improvement. Good housing

Chapter 2 Theoretical Framework and Literature Review

should also represent the advancement of new housing design, new technology and future housing development.

Technological characteristics. New technology should involve housing and delivery of basic services to the people. New technology should represent human development and exploration. It could build the ecological and intelligence houses, to fully satisfy people's needs.

Sustainable characteristics. The housing sustainable development is the target regulation for future housing construction. According to the "International Living Condition Standard", the housing should make good use of the land, water, space, air and natural resources. The sustainable characteristics are one of the most important attributes in the "International Living Condition Standard" (Zhou, 2000).

Based on the "International Living Condition Standard" and the real situation in Beijing Fengtai District, the researcher organizes and forms eight living condition attributes in Beijing Fengtai District:

Items:

a. Size of housing

b. Size and number of rooms

c. Space requirements, kitchen, living room, toilets, balcony etc.

d. Housing facilities such as electricity, water, Internet, communications, security etc.

e. Recreational facilities

f. Price (affordability)

g. Household members (overcrowding).

h. Other acceptable elements like supermarket, train station and post office etc. these service facility near the houses

These eight living condition attributes could fully determine people's living condition. Also it is supported by many previous researchers (Asami, 2005; Bian, 2005). These eight living condition attributes could fully represent the housing characteristics mentioned earlier (Overcrowding, housing privacy and housing facility). Without good housing characteristics, people's living condition will further be affected. Like the size of housing, size and number of rooms, household members, they could represent that the houses are overcrowding or normal. Therefore, due to the overcrowding problem, people's housing privacy will also be affected. The space requirements, kitchen, living room, toilets, balcony, recreational facilities and service facilities near the houses could represent that housing facilities are very important to our everyday life. However, as the rural migrants cannot obtain the *hukou*, they could not purchase houses in Beijing. Therefore, the housing price is not seriously related to the rural migrants' living condition in this research.

Figs. 2.3 and 2.4, compare the average floor areas of owned and/or rental units in the US, Japan, Austria, Italy, France, Germany, the UK, South Korea, and China. The data indicate that the average floor area of housing units in China is much smaller than those of the other eight countries. Although the data were collected in different years, the comparison has managed to establish one conclusion: the average housing size in China is the smallest. Thus, as the average housing size is very small and with large population in China, it causes a serious overcrowding problem, and further affects people's privacy.

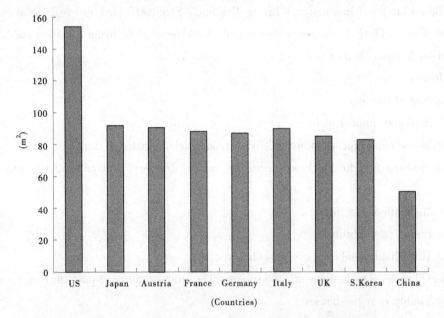

Fig. 2.3　Average housing size(m^2) in selected countries
Source: *Stephen W. K. Mak, Lennon H. T., Choy & Winky K. O. Ho* (2007).

Every woman, man, and child has the human rights to secure a place to live, which is fundamental to living in dignity, to physical and mental health, and to the overall quality of life. The human rights to housing are explicitly set out in the Universal Declaration of Human Rights, the International Covenant on Economic, Social and Cultural Rights, and other widely – adhered international human rights' treaties and Declarations. Despite the widespread recognition of the human rights to adequate housing, the UN Center for Human Settlements estimates that over 1 billion people worldwide are

Chapter 2　　　Theoretical Framework and Literature Review

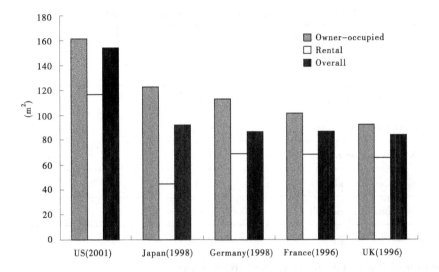

Fig. 2.4　Average housing size (m^2) in selected countries by tenure
Source: Stephen W. K. Mak, Lennon H. T.. Choy & Winky K. O. Ho (2007).

living in inadequate housing and even worse, 100 million are now homeless (Zhou, 2000).

2.5.3.2　International Living Condition Features

Human living conditions have changed a lot over thousands of years, whereby from people living in a shabby shelter to a nice bungalow, this transformation has been common to people all over the world. At this junction, some international living condition features have begun to be established, which have been made the targets for future housing development (Zhou, 2008).

Standard One: The more appropriate living area

Living area refers to the housing size in which people live. Good living area should have enough living space for everyone. According to the investigation by "Assessing Global Housing Conditions": in some rich countries where the average income is more than 10 000 dollar per year, the average living area is more than 20 square meters and some countries like the US have their living areas reaching 60 square meters; while in some countries the average income lies below 5 000 dollar per year, and thus, the average living area stays below 10 square meters. Here is a chart used to elaborate the different incomes that people in different living areas earn in various countries (Gao, 1993) (see Table 2.5).

Table 2.5 Average living area in different countries

Average income in different countries	Average living area (m^2)	Household size
Low – income countries	6.1	2.47
Middle low – income countries	8.8	2.24
Middle income countries	15.1	1.69
Middle high – income countries	22.0	1.03
High – income countries	35.0	0.66

Source: *International housing policy and management*, Gao Peiyi, 1993.

Standard Two: The more perfect living function

Good housing function should include the following factors:

(1) Independence and privacy;

Good house should not be affected by other houses and must protect people's housing privacy.

(2) Convenience and safety;

Good housing function must provide people with comfortable life and good security.

(3) Modern facility;

The housing facility should be more advanced and should be accompanied with high technology.

(4) Health and sanitation;

The houses must be clean and very healthy.

Standard Three: The more completed living facilities

(1) Kitchen and toilet modern facilities where they have good water pipe system;

(2) Good electricity system;

The electricity system should be very safe and fully meet the requirement. Also, it cannot always be subjected to power failure.

(3) Good natural gas system.

The Ventilation system must be very effective and keep the air clean for 24 – hours inside the houses.

Standard Four: The more scientific architectural construction

(1) The architectural construction should be in accordance with the international standard, especially with respect to the light, sound, thermal, anti – earthquake, anti – fire, structure safety etc. ;

(2) Use high technology and new construction materials;

(3) Consider about the economy and changeable situations/policies.

Standard Five: The more appropriate architecture type

According to the international housing standards, the following housing types apply:

(1) Housing type should suit the city culture and economy;

(2) Housing type should suit the natural geography environment;

(3) Housing type should be new and represent the housing development.

Standard Six: Good macro – environment

(1) Good infrastructure, like built near to the hospital, gymnasium, school, post office, police station, supermarket, etc. ;

(2) Good public transportation with the inclusion of more parking places.

Standard Seven: The more civilized community environment

The community environment is very important to residents' mental and physical health. Here, the international housing standard helps to create a more civilized community environment.

(1) Public space. Public space is a place for residents to chat and enjoy being together, so the living space should be equipped with some public spaces.

(2) Green community. A community should have some green areas to offer the people the sense of relaxation; also the green areas could improve the natural environment, like providing cleaner air and stopping soil erosion and so on.

Standard Eight: Physical and mental function

The housing design should have the functions to improve residents' physical and mental health.

(1) Physical function. The housing design should have the physical function like sunshade, soundproofing, solar energy use, day lighting, ventilation, security etc.

(2) Mental function. Improve living condition, privacy and independence; provide different relaxing areas for people of different age groups.

Standard Nine: The more comprehensive social service

International living standard refers to a social service which is not just a common community service like cleaning floors, grass cutting etc, and social service should be more comprehensive, more profound like family and social services, for example solving family trouble and helping disabled people etc. (Gu, 2002).

There are nine housing standards that are included under the definition of good living conditions. To get good living condition, these housing standards should be met and usefully carried out in the future housing development.

2.5.3.3 Summary

Living condition has universally undergone a lot of improvement. Also, with the improvement, it has left a great deal of impact on the society. Living condition not only changes the people's schedule and way of working, but it also affects the quality of life. In the entire world, the living condition should become better and better in the future. Good living condition should have good housing characteristics inside and around the house; also, the comprehensive social service should also be included. Therefore, to improve the living condition, good housing characteristics play very important role in human's civilization.

2.5.3.4 Living Condition in Some Developed Area

a. Living condition in Hong Kong

In Hong Kong, the land is very sparse. Hong Kong has one thousand square kilometers, where 80% of them are mountains. In some high – density areas one square kilometer can be accommodated by ten thousand people. Therefore, living condition stands to be a big problem in Hong Kong.

From 1950, Hong Kong began to take steps to solve its housing problem (Wu & Zhang, 2002). The three – step resolution is:

Step One—Public housing plan. This step was carried out from 1953 to 1972, mainly to solve the housing living problem for poor people. The characteristic is to build a lot of apartments with basic housing facilities. These dwelling places were not for sell, only to be rented out to the poor people. The people only paid little money and they could live in the house for decades. Until 1970, almost one 1 500 000 had been accommodated in these public houses. This plan had solved the housing problem and made the society more stable.

Step Two—Ten years housing construction plan. This plan took place in the period of 1973 ~ 1983 and this was the turning period of housing in Hong Kong. The main target is to build more settlements with good facilities. These houses have independent kitchens and toilets, and they were also equipped with better facilities than before. This plan is to make sure everyone had places to live in. Although these housing areas were small, this was already a big improvement in Hong Kong.

Step Three—Long – term housing policy. This plan has started from 1984 until now. The main target is to change from providing people with low – cost houses to encourage people to buy middle and high cost ones. The policy automatically changes from welfare housing to commercial housing.

Now the housing status in Hong Kong has not been satisfactory. The main housing area uses 51 700 305 square meters, where 3 317 000 people live in these settlements.

Chapter 2 Theoretical Framework and Literature Review

Therefore, the average housing area in Hong Kong is 15.6 square meters (51 700 305 ÷ 3 317 000), less than British (21.34 square meters), also less than Japan (18 square meters) (Yi, 1997).

In Hong Kong, the housing living standards have the following items:

(1) Good housing facilities;

The house should have good facilities like a kitchen, bathroom and balcony etc to meet people's needs.

(2) Housing height;

The floor height should be high and cannot less than 2.7 meters.

(3) Day lighting and ventilation;

The houses cannot be dark in the daytime and it should have good air ventilation.

(4) Fire safety standards;

Need to prevent the fire and have a good fire prevention system.

(5) Environmental standards.

Good environment make people happy and They will enjoy their lives more.

All of these standards are very strict and well-operational. Now the housing facility in Hong Kong is one of the best quality in the world, the only drawback is the high housing price and small housing area. To increase the average housing area is a critical problem and it needs to be resolved in the future.

b. Living condition in Japan

After World War II, Japan was self-reliant in solving its housing issue. Because of the catastrophic war, 4 200 000 homes in Japan were destroyed. Thus, the Japanese government and local council had worked together to solve the housing problem. After the World War II until now, the Japanese central government and local government have all been concentrating on building public housing to solve various housing conflicts. Japan has built a lot of public houses for the low-income families, where the housing area is 70~90 square meters. As the housing price is very low, the public housing has solved a lot of problems for the poor people. From 1955 to 2000, Japan had provided 1 500 000 public houses for the low-income people (Wu & Zhang, 2002).

Most of the people in Japan, much like people in the rest of the countries, would want to purchase private housing, with good housing facilities and surrounded by good environment. For the housing location, the majority of the people want to purchase houses in the sub-urban areas. Here are some reasons for the people to prefer to purchase houses in these particular areas.

(1) In the urban area, the housing price is high, so a lot of people want to purchase the housing in the sub-urban areas instead.

(2) Because of private car popularization, also because of good transportation. The traffic is not a problem in Japan. So a lot of people would like to choose to live in the sub-urban areas.

(3) Good natural environment, like fresh air, clean mineral water, quiet, green area etc. Also the infrastructure is very good in Japan.

(4) The living condition tends to suffer in the urban places where i they are very densely populated. Also the pollution and noise infecting the urban area have compelled the Japanese to live in the sub-urban areas.

The housing characteristics—Tradition and Intelligence tend to be integrated in Japan. Here, when the architect designs the house, first the architect will try to maintain the Japanese culture, and then the architect will mix the culture and intelligence together. Thus, the housing in Japan is a representation of the Japanese culture.

Standard Living Condition in Japan

In Japan, the residents pay attention to seven key factors:

(1) Construction;

Good house must show construction with high quality as it contributes to people's satisfaction.

(2) Facility;

Also the housing should have good housing facilities to meet people's needs.

(3) Indoor environment;

The house should be clean and have a good sanitation system.

(4) Indoor decoration;

The interior part of the house should have some decorations.

(5) Service system;

The service system should be good around the house and provide convenience to people.

(6) Natural environment;

The natural environment should be good around the house.

(7) Social environment;

The security system should be good to protect people.

(8) Affordability.

The housing price should be affordable.

(Zhou, 2000)

Good living should include these eight factors. Housing should have good quality, seeing that in Japan, earthquake is happening common occurrence. The Japanese always stress on the housing facility, natural and social environment, service system and indoor decoration, where all these factors do affect people's living condition.

In Japan's "21 century Housing Development Plan", some aspects are highlighted:

(1) Residents are welcome to participate in the housing design;

(2) Good function, comfortable, high standard decoration material and good facilities;

(3) Good use of the housing resources system;

(4) Good construction material and facilities.

Residential performance standards in Japan, further lists down these aspects:

(1) Structure safety;

The housing structure should be strong for people to live inside safely.

(2) Fire safety;

The good fire prevention system must be assured to protect people from fire or fire - related incidences.

(3) Durability;

The house could be used for a long time and furnishing and facilities inside are in safe condition.

(4) Maintenance convenience;

The houses need to do maintenance every year and maintain good condition.

(5) Energy conservation;

The houses need to save energy and be equipped with environmental protection.

(6) Ventilation;

The houses should have a good air ventilation system.

(7) Day lighting;

The houses should have sufficient light in the daytime and receive good exposure of the sun.

(8) Sound insulation;

People should be able to live in peace and not distracted or disturbed by any external noises.

(9) Barrier - free design.

Some housing facilities should be designes for the disabled, like the slope etc.

(Shen, 2006)

In Japan, the earthquake and typhoon are somewhat frequent, so ensuring structural safety is crucial. Also other factors like fire safety, durability, maintenance convenience, energy conservation, ventilation, day lighting, and sound insulation are necessary to ensure good living condition. Good housing should include all these factors. In Japan, the barrier - free design is very good for the disabled people. These barrier - free designs could help the disabled people live and work just like other normal people.

Overall, In Japan, the housing living condition is very advanced. Japan has a very humanized housing policy; therefore the houses are very practical. Also the housing facilities and environment in Japan are very advanced. The only drawback is that the land in Japan is very scarce and limited, causing the increase in the housing price and the housing areas to be crowded. For the future, the Japanese government tries to solve these problems by building more houses and improving the living condition; not only that there are tasks to fulfill, but also they serve as a challenge to accept.

2.5.3.5 Living Condition Development in Future

Affected by the "Sustainable Housing Development" idea, the majority of the countries' housing development systems undergo three stages: Energy conservation and environmental protection stage, ecological stage and comfortable and healthy stage. At the moment, the stage that seems to be prominent is the stage of trying to increase housing's level of comfort and trying to promote healthy living among the citizens. Before this, in the housing development, a lot of housing concepts had already been introduced. The developed countries had done a lot of research before and some very valuable findings have been derived.

a. Energy conservation housing

The energy conservation housing concept put forth in 1970 (He, 2005; Xiang, Y., 2003) had started with the fact that a lot of countries tried to reduce the housing energy waste especially in the heating, air conditioning, hot water supply, cooking, lighting, household electrical appliances by employing various strategies,. Developed countries, to add, have used a lot of new technology in housing and energy – saving.

(1) UK. In the UK, the energy conservation concept began from 1986, and fortunately, it had been fruitful. What they did was using the local housing material and reducing the transportation fee. Also they worked on decreasing the housing cost. Inside the house, the low energy consumption facility was used. As a result, the energy conservation in its housing decreased the energy 75% more than before;

(2) Sweden. In Sweden, the houses use a lot of recycled materials for heating, and concentrated heating is very common in Sweden. Its housing heating uses 80% of the industrial wastes;

(3) The US. In the US, the housing construction uses high insulation material and new technology. The material could keep warm in the winter and keep cool in the summer. So, every year, the dwellers could save 30% energy in housing;

(4) Germany. In Germany high technology is used to build housing and reduce 30% energy waste. In Germany, the solar system is used very efficiently. The government also

encourages the use of recycled energy.

b. Green ecological housing

Green ecological housing is a very popular concept in the world (Yu, 2003). Ecological housing is a mixture of normal housing and natural environment together. It could balance the indoor and outdoor environments and form a circumstantial system. The green ecological housing emerges, marking the fact that the housing development is off to a new start.

(1) The US. In the US, the green housing is very mature. Almost all new houses use this green concept. In some areas, the power, heating and cooking all depend on the solar energy. The water could be recycled. Because of the natural environment's circumstantial system, this kind of housing not only can save energy, it can also benefit people in terms of improving their physical and mental health;

(2) Canada. In Canada, the green building concept is applicable to the office buildings. In the office green buildings, the toxic gas from the decoration could be reduced and it is very good to people's health. Also the power water and construction material are obtainable from the natural environment. The green housing also enables the recycling of water. Later, the Canadian government has expressed its desire to use this green housing concept in the residential housing.

c. Intelligent housing

The first intelligent housing emerged in the US, then Japan, British, Spain etc (Cai, 2001; Mitchell, 1996). Intelligent housing not only changes the way of life, way of working, but also it improves the housing development, intelligent communication, housing service, also the traditional home appliances.

(1) US. In 1984, the electronic and computer technology were used in the first intelligent building. Inside this building the air-conditioner, water supply, electricity supply, fire protection system were all controlled by the computer. Also, the central computer could automatically perform the analysis of the telecommunication, email, marketing, scientific calculation etc. Later, the office buildings like IBM, DEC were all made into intelligent buildings;

(2) The UK. The intelligent housing in the UK has grown early and developed rapidly. In the UK, almost all the new terrace houses use the intelligent housing concept. These houses made by natural and local materials are therefore very ecological. Also these houses can save the natural resource when they are automatically controlled by the computer system. Water could be automatically recycled and the security system could detect whether some people have entered or exited. Even in cases where the people sleep, the computer can automatically adjust the temperature and humidity—all for the sake of

comfort and convenience.

d. Barrier – free housing design

Before talking about the energy conservation housing, green ecological housing and intelligent housing, these housing types have all translated the cutting – edge features into mere technology. However technology is not the means in itself—human, especially the old and disabled people should be understood and respected (Chen, 2003; Michael, Samantha and Emily, 1998). The barrier – free housing represents this need and the old and disabled could benefit from this type of houses.

(1) The US. In the US a lot of houses are especially built for old people, to provide them with ongoing assistance and medical care. There are three types of houses constructed for this purpose: independence – living, assisted living and acute – care. In the barrier – free housing, a lot of facilities are made especially for these disadvantaged people. The barrier – free housing is supported and encouraged by the US government;

(2) Denmark. In Denmark, the barrier – free housing resembles the common housing at the outside, but inside the kitchen, toilet, living room and bedroom, even the staircase all use the special barrier – free design. It guarantees that each house has 67 square meters, and it must have 24 – hour emergency communication. In Denmark, the barrier – free housing has been a very successful project.

e. Conclusions

The above house – types have all focused on "human, nature, technology", and the main target is to increase the housing, as well as the living conditions. Every individual needs adequate housing to live in, so the government should pay attention to the housing development, use of good technology and formulation of good housing policies, so the residents can enjoy and benefit from good living conditions.

2.5.3.6 Discussions and Conclusions

From the globalization, in the developing and developed countries, the housing and living conditions have greatly improved. The housing affordability and private housing may increase; same goes with the people's average living area, and the housing environment will improve. During this process, the provision of some common characteristics has been found to be problematic.

Firstly, house – purchasing is an important thing for most people. Thus, the housing quality and performance will affect people's lives. In the developed countries, some laws and regulations have been established to guarantee housing quality and protect consumers' interests. However, in some developing countries, the housing laws cannot answer for the atrocious housing quality, like noticeable cracks and collapses that frequently happen. Learning from the developed countries' experiences and improving the housing quality is a major task for the developing countries.

Secondly, the living condition improves a lot too within these years. All the countries have to undergo some obligatory processes in the housing development namely the energy conservation and environmental protection stage; ecological stage; also comfortable and healthy stage. The main target is to bring about people's comfort and health. Therefore, improving the living condition is the main idea.

Thirdly, after analyzing the housing living condition in developed countries, it is found that the improved housing and living condition are all based on the objective, not subjective views. They have never taken into account human's own perspectives. This is the drawback witnessed in the housing development. In future, the residents should participate in the housing design and construction, and only then the living condition can be improved quickly, practically and more effectively.

Chapter 3
Housing System: the Chinese Perspective

3.1 Introduction

The aim of this chapter is to examine the housing system in China in relation to the country's economic growth and housing for migrants, especially rural migrants.

This chapter will first look at the economic performance in China and its effects on the housing system, before discussing the history of the housing development in China. Some examination of housing policy development starting from China's established year will also be conducted. Finally, the chapter will discuss the housing policy for migrants, especially for rural ones residing in Beijing. The discussions are supported by contextual findings from the fieldwork.

3.2 Chinese Economic Performance

Since 1949, the industry was scarce and production had been very low. The economy was supported by agriculture which became the major contributor towards the growth of the country as a whole. During this hardship period, the country had suffered from the Japanese invasion and domestic war that China had become almost wholly – dependent. After it's establishment in 1949, China practiced a relatively open economic system driven by human basic need—food. The government encouraged people to contribute to agricultural production to decrease starvation. During this period, the industry had a slow progress.

From 1978, the government adopted an aggressive open economic system driven by market force. During this period, the economy increased very fast and both of the industry and agriculture had had remarkable improvement. The Gross Domestic Product (GDP), following a growth of rate 9.8% in the 1980's, grew at an annual rate of 10.4% in the 1990's and continued to enjoy a growth rate of more than 11.0% during the period of late

1990's until 2007 (Yang, 2000).

Gradually, the country moved from its role as a producer of agriculture to the hub of agriculture and industry both as a result of the open economic system to import and export products. The government also played an active role to attract investors to develop the manufacturing industry. Relatively speaking, in the 1990's, the industry and electronic manufacturing had experienced a tremendous growth over the decade. Because of the policy shift, the GDP of the national income increased from $ 6.291 billion in 1982 to $ 247.0 billion in 2006 while the manufacturing industry had a tremendous improvement and ranked 4th in the world, and the high technology electronic manufacturing ranked 2nd in the world (see Table 3.1) (Zhang, 2010).

Table 3.1 Gross economic development in China

Sector	GDP by year ($ Billion)	
	1978	2007
Agriculture	2.692	94.197
Industry	1.1175	74.221
Manufacturing	1.170	37.350
Construction	0.780	33.760
Others	0.531 5	7.472
Nation income	6.291	247.0

Source: *Housing and finance renovation in China*, Zhang En Zhao, 2010.

There are three factors which influence the economic growth. Firstly, the government had paid attention to, and prioritized, the agricultural productivity. Because of the agricultural productivity increase, there was enough food for the Chinese and starvation was able to be eliminated. Secondly, the growth of the industry and manufacturing sector which could produce more products to sell domestically and internationally also mirrored the government's open economic policy. Finally, because people's standard living condition had increased, more housing was needed. It influences the construction sector to build more housing with good housing features and characteristics. More and more investors had invested money in real estates and intensified the construction in China, evident from 0.780 $ Billion in 1978 to 33.760 $ Billion invested in 2007 (Zhang, 2010). The whole housing development in China had a steady growth of GDP shares within the period.

China's rapid economic growth also had an incisive influence on the migration. More and more peasants who worked in the agricultural sector before were willing to migrate to urban areas and searched for jobs there. Some people from small cities would like to migrate to big cities and work there. According to the 2009 population census, the number of urban and rural migrants in Beijing had reached 7 million; two thirds of the migrants were from the rural areas. Therefore, so many migrants had been affected and positively, they helped boost the industrial, manufacturing and construction sectors. Thus, it is only necessary that the government pay more attention to the secondary and tertiary sectors, as far as economic activities and migrants' employment are concerned.

3.3 The History of the Housing System Development in China

3.3.1 Elementary Welfare Housing System in China (1949 ~ 1980)

When China was established in 1949, the national economy was very poor. During this time, only big cities like Beijing and Shanghai had a few high – rise buildings, and the middle and small cities had single – storey houses. People live in low quality houses often face the non – existent basic housing facilities, housing humidity, leakage, and also poor ventilation. Particularly during the heavy rain, many houses in low – lying places have had to deal with flooding.

In this period, the government had tried to solve the housing problem for residents and used a new housing system in China. The welfare housing system started from 1949 and had been implemented all across China. The welfare housing system refers to the housing that is constructed, managed and maintained by the government department or company, where the government will provide free land. The housing provided to residents for free, with only a little housing rental charged. In this period, housing was not a product; it was merely a form of welfare to serve people. In a closer analysis, the welfare housing rental only charged 0.257 RMB per m^2 per month in Guangzhou in 1952, and it only occupied 9.31% of family's total monthly income. The prime minister Zhou En Lai has stated that the housing should serve more people and give residents more welfare. Therefore, during this time, the welfare housing rental decreased every year. In 1953, the welfare housing rental occupied 8.7% of the family monthly total income; in 1958, the welfare housing rental decreased to occupy only 8.41%. The government tried to help more poor people to live in the housing with good facilities, but only with little fee. The lowest welfare housing rental in that time was only 0.12 RMB per m^2 per month in China (see Table 3.2)(Yang, 2012).

Table 3.2 **The percentage of welfare housing rental among total family income in China**

(%)

Year	The percentage of welfare housing rental among total family income
1949	9.13
1953	8.7
1958	8.41
1963	7.0
1970	1.2
1975	1.1
1980	1.04

Source: *The record of welfare housing system in the elementary stage of China*, *Yang Ji Rui*, 2012.

From 1949 to 1980, the welfare housing system in China had been carried out very well and solved a lot of housing problems, at least all the poor and homeless people had had the chance to have a place to live. Everyone in that period all appreciated the good socialism and the Chinese government helped them to reside in decent housing. Meanwhile, although the welfare housing system in China had been smooth, there were still some drawbacks to this housing system. The welfare housing was first constructed, managed and maintained by the government department or companies, so it gave a heavy economic burden to both the agencies. During that time, China was just established and still very poor. In some places, the government and the companies did not have enough money to build the welfare housing thus the construction process was very slow; some even stopped altogether. Therefore, the welfare housing was very inadequate in that period.

Secondly, the welfare housing management system was not rational in that period. As stated before, the housing rental only occupied a little percentage in the family total monthly income. People's salaries increase every year in China, while the housing rental did not increase for more than 30 years. In specific, the housing rental in that period was only 0. 2 RMB per m² per month in China and it did not increase for a long time (Wu & Luo, 2009). The housing rental was too little even not enough to cater for the maintenance, so the conditions of welfare housing in a lot of cities was very poor.

Thirdly, in that period people had depended too much on the welfare housing. As the government provided good housing for each resident, a lot of people had become lazy to work and they also did not take care of t the places in which they lived. Many residents were not willing to do housing maintenance because they took for granted that the government or company would do that for them. Thus, the welfare housing system had made many people idle, and housing maintenance was easily neglected. This made the government and company's burden exacerbates (Xiao, 1993).

Finally, the welfare housing system created was unfair. In that period, the welfare housing was provided by both the government and company. However, in some poor places, owing to the bad finance, the government and company had no capability to build the housing. What happened was, very few people could get the welfare housing, while some people had obtained welfare housing by means of corruption even through fights. If a family had many children but only one was entitled to the welfare housing, the house would be given to the boy. At that time, many people feel that the welfare housing system was not fair, and therefore the housing system improvement was demanded.

3.3.2 Elementary Commercial Housing System in China (1981 ~ 1985)

Because China had been executing the welfare housing system for more than 30 years, although the housing problems were solved for many poor people, it had created a heavy burden to the government and company. Thus, it can be said that the national economy and housing development were obstructed by the welfare housing system.

Starting from 1980, the Chinese government wanted to solve this housing problem and created a new housing system to be used in future. In 1980, the Prime Minister Deng Xiao Ping mentioned that China's housing system should be open, and the housing purchase and sell could also be used. Then, the Chinese government decided to try to use the commercial housing system and tested it in four cities only namely Zhengzhou, Changzhou, Siping and Shasi (Luo, 2007). Until 1984, the commercial housing system was carried out smoothly in the four cities, so the Chinese government decided to expand the commercial housing system to the whole country. During this period, the government encouraged the residents to purchase the commercial housing and stop the commercial construction in the future(see Table 3.3).

Table 3.3 **The average housing size of each person of urban population in China from** 1980 **to** 1985

Year	The average housing size for urban population in China from 1980 to 1985 (m^2)
1980	3.6
1983	4.6
1985	6.0

Source: The national housing investigation report in China, 1986. CCG, China Central Government, 1986.

The commercial housing system has some drawbacks. Despite the fact that the commercial housing needed purchases by the residents, in 1980's, people's salaries were very low and it was still very difficult to afford the commercial housing. Then, the government and company decided to afford two thirds of the commercial housing price, with the people paying one third of the price. Even though the government and company paid for the most of the commercial housing price, people in that period still found it difficult to afford the one third of the price of the commercial housing. Thus, until 1984, the commercial housing only sold 20% of the total commercial housing, while the other 80% commercial housing still could not be sold out and were subsequently kept by the government and company. The government and company's economic burden continued to be draining for them (Huang, 2010).

Also, there were some problems of the commercial housing distribution system (Wu & Luo, 2009). Firstly, some people who had purchased the commercial housing earlier could enjoy a lot of advantages, while some people who had purchased the housing a little later was denied the advantages. Thus, many people were not satisfied with this commercial housing system. Secondly, for the people living in the urban areas and holding the urban *hukou* (household registration system) they could enjoy more advantages granted by the housing system, while the people living in the rural areas and have the rural *hukou* cannot enjoy these advantages. Finally, for the people who had lived in big houses before, in the case if they wanted to sell the housing and then purchase the commercial housing, they could enjoy more benefits granted by the government or company. However, for the people who lived in the small housing or even without any house previously, they could enjoy only a little allowance around 1 ~ 3 RBM per month or even without allowance. So many poor people had filed complaints about this commercial housing policy and they demanded that the policy should be changed.

This was the elementary stage of the commercial housing system and as it had a lot of problems, it should be improved. From 1980 to 1985, the commercial housing system was tested and the Chinese government decided to improve it in 1985 (Luo, 2007). Therefore, the commercial housing system could be improved and operated well in the next development stage of the housing system.

3.3.3 Commercial Housing System Improvement in China (1986 ~ 1993)

From 1986 to 1993, the commercial housing system was the second improved housing system in China. In the 1980's, the national economy had increased rapidly thus the government decided to improve the housing system. During that time, a lot of people were still poor thus the majority of them still had to settle in the welfare housing. The government wanted to carry out the commercial housing system, so the welfare housing must be transferred to the commercial housing.

During that time, people lived in the welfare housing with only a little housing rental to pay. The housing rental was too little that the payments were not enough to support the maintenance and new construction underway. To improve the commercial housing system, together with the transfer from the welfare housing to commercial housing, the government decided to increase the housing rental. In effect, from 1986 to 1988, the welfare housing rental had increased. Fortunately the residents' salaries had increased a lot so there were not too many complaints reported then (Xiao, 1993).

Also to decrease residents' economic burden to purchase commercial housing, the government published a new policy during the years of 1988 ~ 1993. If the residents sought to purchase commercial housing, the government, the housing company and individual would each pay 1/3 of the total housing price (Xiao, 1993). This new housing policy had motivated a lot of high and middle income people to purchase the commercial housing. Thus, during 1986 ~ 1993, the commercial housing system had improved and had been well implemented in China.

While the commercial housing system improvement observed during 1986 ~ 1993, there were still some advantages and disadvantages of this housing system. There were two advantages of the commercial housing system. Firstly, residents' housing quality has increased. With the new housing construction improvement and real estate development, more and more people had paid attention to the housing area and housing quality.

Therefore, with the housing construction improvement and real estate development, the average living area of each person in China has increased (see Table 3.4). It represented that the Chinese residents' quality of life has increased. But compare to the developed country like US, the average living area in China is still very small. Until 1993, the average living area in US almost 4 times bigger than China and it represented that the US residents' quality of life is much better than China (Xie, 2009).

Table 3.4 Average living area in China and US from 1986~1993

Year	China (m^2)	US (m^2)
1986	9.4	36
1987	12.7	40
1988	13.0	—
1989	13.5	—
1990	13.7	—
1991	14.2	48
1992	14.8	56
1993	15.2	60

Source: Real estate development in China (Xie, 2009).

Secondly, more and more Real Estate Company had mushroomed that the housing construction had quickly developed. Until 1992 the housing construction had come to its peak. In some parts of China, the national economy could occupy 1/3 of the housing construction. In the Hainan province, for instance, the Real Estate Company had reached 400 and the housing construction in Hainan reached 755 700 000, 1/3 of the national income was obtained by the housing construction area. Thus, the commercial housing system contributed greatly to the national economy (Wu & Luo, 2009).

However, at this time, the commercial housing system had two disadvantages (Xiao, 1993). Firstly, even if the government, company and individual paid 1/3 of the total housing price, some very poor people still had problems paying for it. Also, the welfare housing system was already terminated and the housing rental increased. Thus, a lot of poor people became homeless and they had to live in the shelters. During that time, a lot of poor people complained that the commercial system was not fair and they only served the rich people.

Thirdly, because the government and company each needed to pay 1/3 of the total housing price, thus the government and company's economic burden was still too much to bear. Also, in some poor places the local government and company' finance was not good and they did not have enough money to purchase houses for the employees. Sometimes the housing purchasers needed to wait a long time for the government and company to approve the housing payment, so the commercial housing system did not carry off well in some poor areas.

Overall, the commercial housing system improvement as adopted by China suggests that China was looking for the most suitable housing system for its people. The commercial housing system was seen as a stepping stone for the next stage of development of the housing system in China.

3.3.4 Housing System Exploration and Legal Reserves of Housing Acquisition System in China (1994~1997)

Before 1994, Chinese government was looking for the best housing system that could be adopted nationwide. There were two housing systems that were used before: the welfare housing system and commercial housing system. The two housing system have both the advantages and disadvantages, but in 1994, the Chinese government found and published a new housing system—legal reserves of the housing acquisition system. This housing system was much more scientific than the previous two housing systems and it could carried out smoothly, as a fact, this housing system has stood to become the formal housing system in China and is used until today (Yang, 2012).

In July 1994, Chinese governments published the legal reserves of the housing acquisition system aiming to support and help the middle and poor income people. This housing system was carried out throughout China and it was highly welcomed by the middle and poor income people.

The legal reserves of the housing acquisition system had started from 1994, where the government and company would take 5.0% of the salary from people's accounts and deposits in the bank (Wang & Wang, 2002). The money was kept and controlled by the government and interests paid to residents. With the salary increase, the government took more portions of income to reach 12%. If the residents needed to buy their houses, they could withdraw all the money that was deposited before, also they could apply for a loan from the bank with very low interest. This housing policy mainly helped the middle and poor people to purchase houses, also it provided a loan for residents to purchase houses.

According to the China Housing Investigation in 2005, this housing policy encouraged and affected the urban residents to purchase houses from 5.3% in 1995 to 66.61% in 2005(see Table 3.5)(Wu & Luo, 2009).

Chapter 3 Housing System: the Chinese Perspective

Table 3.5 China housing investigation in 2005

Housing type	Urban residents (%)	Rural residents (%)
Self – built	10.99	93.82
Purchase housing	66.61	2.03
Rent housing	18.82	1.61
Others	3.58	2.54
Total	100	100

Source: Housing renovation in China (Yang, L. & Wang, Y. K., 2006).

From the Housing Investigation in 2005, it is shown that the legal reserves of the housing acquisition system encouraged and increased residents to purchase housing in urban areas. Meanwhile in 1995, only 5.3% of residents had bought houses in urban areas. Until 2005, this number increased to 66.61% and it is still increasing (Yang & Wang, 2006). Because this housing system could provide middle and poor people with loans with very low interests, it had motivated these people to buy houses. Back in 2011, the government increased the legal reserves of the housing acquisition loan from 4.0% to 4.2%. But, it was still much lower than the commercial loan which could be obtained directly from the bank (Yang, 2012).

From 1994 to 1997, Chinese government tried to carry out the legal reserves of the housing acquisition system and they received some impressive results. Until 1997, in urban areas more than 50% of the whole residents had already lived in houses that they purchased by themselves. This housing policy has more advantages than the welfare housing system and commercial housing system (Yang & Wang, 2006).

Firstly, this housing system could encourage the people especially the middle and low income people to buy their own houses. A lot of poor people could work for two or three years, then they would be entitled to apply for the housing loan. Thus, it encourages people to work hard and it makes them more diligent. Secondly, this housing policy decreases the government and company's burden and there is no need for residents to purchase houses. The government and company only provided some incentives for residents and they only focus on their production. Finally, this housing policy forces the residents to deposit money in the bank for residents' economic stabilization. If the residents need to purchase houses in the future, then they could withdraw the money. Also, it is much safer to deposit the money in the bank.

There were still, however, some drawbacks in this housing system. Firstly, the residents need to deposit some money in the bank according to their salary, so the high income people do not have to depend on this housing system to purchase. Meanwhile, although assisted by the bank, the very low income people still have high economic burden to purchase houses. Thus, this housing system cannot cover for the very poor people and therefore, it needs more improvement. Secondly, to enjoy the legal reserves of the housing acquisition system, the residents need to have a formal job. Meanwhile, a lot of people do not have a job, so they cannot enjoy this housing system. In fact, only a small group of residents can benefit from this housing system. Finally, at the initial stage of this housing system, some residents only deposit money in the bank but cannot enjoy it because the system carried out turns out to be very slow.

The legal reserves of the housing acquisition system are much better than the welfare and commercial housing system, but it still needs some improvement. In the next few decades, this housing system is improved and completed and used until now. To benefit everyone, this system targets to achieve good housing system development in China.

3.3.5 Legal Reserves of Housing Acquisition System Improvement in China (1998~2007)

From 1980 to 1997, the housing construction in China created an incredible record and the housing system in China slowly became mature. During the 18 years, the national economy became strong and residents' income had improved a great deal. In 1997, the average resident income reached 5 160.3 RMB, increasing more than 60% than it had in 1980. Also, urban residents' bank deposit reached 4 627.98 billion RMB. With the strong national economy, the residents wished to increase their living condition, especially concerning the housing areas. In 1997, the average living area in the urban areas was 17.8 m^2. So, the inclination to increase the housing areas became strong (Wu & Luo, 2009).

Finally in 1998, the housing system in China had an incredible improvement. The Chinese government declared to improve the housing construction and let the housing construction become the backbone of the national economy(see Table 3.6).

Table 3.6 The investment of real estate from 1998 to 2007 in China

Year	Investment/cash (billion RMB)
1998	430
1999	450
2000	490
2001	600
2002	700
2003	900
2004	1 100
2005	1 400
2006	1 800
2007	2 200

Source: *The real estate of housing investigation in China*, *China ministry of finance*, 2007.

The legal reserves of the housing acquisition system improvement forced the real estate market to develop fast; there were some main points of the improvement.

Firstly, the government tried to come up with a more economically affordable housing that to will be able to help the middle and low income people. The economically affordable housing refers to the housing with low price, mainly to sell to the middle and low income people. To purchase the economically affordable housing, they should have the urban *hukou* (urban ID) and follow all the requirements that are made by the government. During this period, even the economy has increase rapidly but still a lot of poor people cannot afford to have their own homes. Therefore, the government forced the developers to construct more economically affordable housing for middle and low income people. Meanwhile, for the high income residents, they cannot purchase the economically affordable housing and can only buy the high – cost commercial housing.

Secondly, the government will improve the housing business market to be more standardised. With the housing system development and housing business deals increasing, the housing market should be more standardised than before. Therefore, the government will improve the housing market regulations to be more systemised. The residents who want to purchase the economically affordable housing, commercial housing or rent housing should all follow the housing market regulations. Also, the purchasers will be recorded in the government department. All these methods will let the housing purchasers have more confidence in buying houses.

Thirdly, the government will improve the housing financial system. To give more convenience for residents to purchase houses, the government allows all the banks to provide loans for the local housing purchasers. Also, the banks could allocate more time for the accommodators to refund the money. Thus, the residents could purchase their dream homes with more confidence and could have more time to refund the money.

Finally, the government will also enhance the housing tenement management. The housing tenement management company will be located near the housing and it will perform the housing renovation and management. Thus, the housing management will be more standardized and scientific. In the following decades, this housing management system is still used until now, and is highly welcomed by the residents (Wang & Huang, 2011).

The legal reserves of the housing acquisition system improvement created an incredible record and had achieved a lot too. Here are some housing improvements reported in China.

Firstly, urban residents' living condition has improved a lot, and the dwelling concept has become more advanced. After the legal reserves of the housing acquisition system improvement, the urban residents' living condition and living standard have all improved as well. In the urban areas, the average living area continued to increase from 22.8 m^2 in 2002 to 28 m^2 in 2007 and achieved the middle among developer countries. Until 2007, 83% of the urban residents had had their own houses. All these improvements show that this housing system is very suitable to be carried out in China (Feng, 2009). Also, with the science development, the building technology becomes more mature. Thus, residents not only need a shelter to live, but the most important thing to remember is that the housing should be of service to the people. Therefore, the urban residents' dwelling concept must be more attentive to the housing quality, housing environment and housing functions. Overall, the dwelling concept can spur the housing development, and the housing development could further be carried out successfully.

Secondly, real estate has developed well and become the backbone of the national

Chapter 3　　　　　　　　Housing System:the Chinese Perspective

economy. With the housing development in China, the real estate has developed exponentially. In some parts of China, the housing investment could occupy 50% of the local economy. In some big cities like Beijing, Shanghai, Shenzhen and Guangzhou, the housing investment could occupy 60% in the local economy. From 1998 to 2007, the housing investment increased from 500 billion RMB to 3 000 billion RMB (Hu, 2009). Thus, to manage the housing market system, the good housing system must be made a reality. Also, with the legal reserves of the housing acquisition system improvement, the housing investor shift from government to developers. And the investments were all afforded by the developers. The government will control and manage the housing construction and business market. Thus the housing construction and business market will become more mature and will progress well in future.

Thirdly, the housing financial system becomes more mature. With the improvement of the legal reserves of the housing acquisition system, the housing financial system is set to improve and will become more mature. Usually to purchase housing, the residents will get a loan from the bank. Before, the housing financial system was very weak and it was very difficult to get the housing loan. However, now that the housing financial system was much more mature than before, the housing loan has increased a lot. In 1997, the housing loan in China was only 19 billion RMB, occupying only 0.39% of the total loan. However, until 2002, the housing loan in China had reached 800 billion RMB and occupied 7.6% of the total loan. Also, the accommodator could have more time to refund the money (Fu, 2009). It further shows that the housing financial system encourages residents to buy their own houses.

The Chinese government also encouraged developers to do housing investment. Until April 2003, the real estate housing loan reached to 1 835.7 billion RMB in China and occupied 17.4% among total loan. The government also encouraged the Real Estate Companies to do housing business abroad, like Lu Chen China, Hopson Development, Shimao Property and Shanghai Forte Group (Xie, 2009). All these are proof that the Chinese government had paid much more attention to the housing development in China.

Overall, the legal reserves of the housing acquisition system improved a lot during the fourteen years and the Chinese housing system it is still used until now. During these years, this housing system accomplished a lot and benefited many urban residents. Some drawbacks of this housing system persisted, nonetheless. From 2007 until now, the Chinese government will continue to improve the legal reserves of the housing acquisition system.

3.3.6 Legal Reserves of Housing Acquisition System Improvement in China (2007 to Present)

From 1998 until now, the legal reserves of the housing acquisition system has improved a lot, but still some problems remain. From 2007, the Chinese government has worked endlessly to improve the housing system(see Table 3.7).

Table 3.7 The housing construction in some cities of China in 2009

City	Commercial hosing (million m²)	Economically affordability housing(million m²)	Low-rent housing (million m²)
Beijing	28.3	—	—
Shanghai	18.0	—	—
Tianjin	20.0	5.5	0.05
Chongqing	31.15	27.4	0.51
Shenzhen	9.82	0.32	1.4
Guangzhou	9.08~12.0	1.5~1.9	0.08~0.1
Chengdu	10.51	0.207	0.023
Xi'an	15.05	1.55	0.07
Ningbo	3.58	0.02	0.02
Hefei	16.32	0.563	0.08
Wuhan	10.0	—	0.075
Changsha	9.0	10.0	0.1
Zhengzhou	11.0	0.8	0.15
Qingdao	15.0	0.325	0.15
Xiamen	47.0	—	—

Source: The record of housing construction in China, China ministry of construction, 2010. CMC, China Ministry of Construction, 2010.

With regards to the problems that persist, the housing price has escalated too fast which affects the middle income people to purchase commercial housing. The International standard of evaluating residents who have the ability to purchase housing suggests that the ratio of the annual income to housing price should be between 1∶3 and 1∶6. However, starting from 2003, the housing price has started to increase rapidly. From 2004~2007, the housing growth rates were 17.8%, 14.0%, 6.3% and 14.8% (CCG, 2007). The housing price growth rate was much quicker than the salary increase rate. For big cities like Beijing, Shanghai and Shenzhen, the housing price had escalated to more than 10 000 RMB per m^2.

The high housing price growth rates result in a lot of problems and they particularly affect the middle income people to purchase commercial housing. The economically affordable housing constructions are very slow and the poor people are given more priority. Therefore, the middle income people need to purchase commercial housing. However, these people cannot bear the high housing price and they are already burdened even before they purchase any form of houses.

Secondly, the economically affordable housing constructions are too slow. Starting from 1998, the Chinese government has started to construct the economically affordable housing which is the middle low cost housing to help the poor. However, the investment was too low, where it only occupied 4% compared to the commercial housing. Until 2007, there were 10 million poor families in the urban area and the economically affordably housing only solved the housing problems for 0.268 million (CCG, 2007) people. Thus, increasing the economically affordable housing is necessary.

In August 2007, the Chinese government decided to improve the legal reserves of the housing acquisition system again. The housing system improvement mainly sought to solve housing problems for middle and low income people.

Firstly, the government will try to control the commercial housing price and make sure that the middle income people could afford it. This method was carried out in big cities like Beijing, Shanghai and Shenzhen first. From 2007 until 2011, the housing prices in the four cities had decreased slowly. In 2012, the housing price decreased by 20% in Beijing. To control the housing price was difficult, but there was still some achievement reported (Lue, 2009).

Secondly, the government will improve the economically affordable housing investment. Every year, the government will use not less than 10% of the housing investment to embark on the economically affordable housing construction. For the very poor area, the central government will give more subsidies to help promote the economically affordable housing construction (Xie, 2009). Thirdly, the government will give more priorities to provide land for economically affordable housing construction. Finally, the very poor family could enjoy free housing maintenance provided by the government.

Because of the legal reserves of the housing acquisition system improvement, from 2007 until now, the housing in China has shown great achievement. Because the government has paid more attention to the economically affordability housing construction, until 2011, 2.5 million low income urban families' are entitled for the economic housing. Also, the government had tried to control the commercial housing price and helped the middle income family have the ability to purchase commercial housing in urban areas (Wu & Luo, 2009).

3.3.7 Future Perspectives

The legal reserves of the housing acquisition system are carried out smoothly and they show very good results. The Chinese government continues to improve this housing system in China and wishes it could benefit more and more people. In future, the government sets three targets in the housing system.

Firstly, the government seeks to increase the economically affordability housing so that it could improve the low income people's living condition. Secondly, the economic affordability is developed to stimulate the relevant areas of the economic development. Finally, the development of the commercial housing could encourage the consumption thus stimulating the economic development (Bo, 2010).

Overall, the housing system could affect residents' lives, and it also could influence the national economy. With the economy further growing, the Chinese housing policy will become more and more standardised.

3.4 The Housing System in Beijing (Fig. 3.1)

3.4.1 The Housing System for Local Residents

In Beijing, the housing system gives more priority to the Beijing's own local residents. Basically, there are three types of housing in Beijing: commercial housing, economically affordability housing and low - rent housing. The commercial housing and economically affordability housing are not that much different from the other places. Our current work focuses on the low rent housing in Beijing, which is built to benefit the very poor local residents (see Table 3.8).

Chapter 3　　　　　　　　　Housing System:the Chinese Perspective

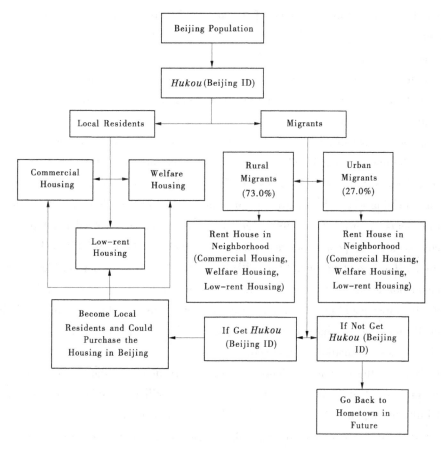

Fig. 3.1　The housing system in Beijing

Table 3.8　Describe the type of housing in Beijing

Housing type	Describe the type of housing
Commercial housing	The commercial housing is developed by developers and sell to local residents, the commercial housing price is controlled by Beijing housing department
Economically affordability housing	The economically affordability housing is supported by Beijing government and built by developers, then sell to local residents with lower price
Low – rent housing	The low – rent housing is supported by Beijing government and built by developers, and then the very poor local residents could rent the housing with very lower price

Source: Housing policy in Beijing, Beijing Government, 2012. BG, Beijing Government, 2012.

The low rent housing is a kind of housing constructed by the government and it is provided for the very poor local residents for rental. The low rent housing is not sold and the housing rental is very low. In Beijing, the family total income is less than 1 200 RMB and with this income, they could apply the low rent housing. The housing rental is calculated by the housing area and one m^2 the housing price rental costs around 2.5 RMB in Beijing (Cheng, 2011).

The low rent housing policy started from 1998 and was only carried out in some big cities like Beijing, Shanghai and Guangzhou etc. Therefore, only the residents from big cities could enjoy the low – rent housing. Until 2007, there were 10 million very poor residents in China who did not have their own houses and they made up 5.5% of the total urban residents. Starting from 1998, the low rent housing has solved a lot of housing problems for the very poor urban residents. Until 2009, a total of 547 000 urban families lived in the low – rent housing (Cui, 2011). Even this type of housing could provide accommodation for the very poor urban families, but the houses were very limited. Also, only the big cities could take advantage of the low rent housing policy, whereas the middle and small cities did not have this kind of policy. Overall, the low rent housing is very helpful and thus, it should be improved.

In Beijing, the commercial housing price is very high and the average commercial housing price reached 30 000 RMB per m^2, and it is still increasing. The economically affordability housing mainly focuses on the middle and low income people, where the price is lower than the commercial housing. Meanwhile for the very poor people and whose income is lower than 1 200 RMB per month they could not afford both the commercial housing and economically affordability housing. Therefore, the low rent housing policy is very helpful for these groups of people. Nonetheless, the low rent housing policy only caters for the local residents only(see Table 3.9).

Table 3.9 Low – rent housing in some big cities of China

Cities(name)	Number of low – rent housing	Number of people apply low – rent housing	Percentage (%) of low – rent housing adopted
Beijing	3 032	14 351	21.1
Tianjin	2 863	26 600	10.8
Shanghai	17 768	45 079	39.4
Guangzhou	2 460	14 914	16.5
Total	26 123	100 944	25.9

Source: The record of welfare housing system in the elementary stage of China(Yang, J. R., 2012).

The low rent housing in Beijing sees the investment of the local government and the local government does both the maintenance and management. In Beijing, until 2012, there were 14 351 local residents applying for the low rent – housing and the Beijing government itself invested 4.744 14 billion RMB for the low rent housing construction and maintenance. In Beijing, the very poor local residents total 1.86 million and a lot of local residents are still waiting for their names to be selected as the rightful owners of the low rent housing (BG, 2012).

For the low rent housing policy, there are many drawbacks in this housing system and thus, it needs to be improved in the future. First of all, the low rent – housing construction is not enough for so many poor people and the government needs to add more houses of this kind. Secondly, the low rent housing system has too much limitation. In Beijing, only income lower than1,200 RMB could enjoy this housing system. However, as it is many people's salary is around 1 500 RMB, and the reality is that they still cannot afford the commercial and economically affordability housing (Wu & Luo, 2009). Thus, the low rent housing policy cannot cover the low income people in a true sense of the word. Thirdly, low – rent housing is difficult to manage. A lot of poor local residents do not take care of their houses and the government ends up having to spend a lot of money on the housing maintenance. Also, many poor people have to use their houses for their small businesses. This is not allowed by the Beijing government (Yang, 2012). Thus, the government needs to hire great workforce to manage these houses. Finally, the local government has exerted too much pressure on low – rent housing investment. From 1998, the central government has increased the low rent housing investment and the investment needed has increased to 49.7 billion per year. All the investment was donated by the local residents. Thus, the local governments have been pressurized by the low rent housing investment.

Overall, the housing system (refer to Fig. 3.1) is carried out smoothly in Beijing but it still needs to improve. The low rent housing in particular, needs to improve so that more residents could enjoy it. All in all, the housing system in Beijing appears to only favour the local residents. For the migrants, rural and urban migrants, they cannot enjoy any housing system in Beijing without the Beijing *hukou* (BG, 2012). The only way for these migrants (rural migrants and urban migrants) is to rent houses that they can find in their neighborhood.

3.4.2 The Unfair Housing System for Migrants

In Beijing, the housing system seems to favour the local residents. This puts urban and rural migrants at the backseat, as they cannot enjoy any benefits from the Beijing housing system.

In Beijing, the local residents could enjoy three kinds of housing. Commercial housing, economically affordability housing and low – rent housing are three kinds of housing that are provided for local residents with different income levels. Meanwhile for the urban and rural migrants, they cannot enjoy these types of houses. According to the previous researches, 95% of the migrants reside in the Beijing rented housing in the neighborhood, and only a few migrants live in the housing provided by the company (Wang, 2007; Wu & Wang, 2002; Pang, 2003). The migrants cannot purchase the commercial housing because they do not have the Beijing *hukou*. Also they cannot enjoy the economically affordability housing and low – rent housing. Due to the fast increase of the population and that so many migrants surge into Beijing every year, there is simply not enough land to build houses and resources are limited for these migrants. With this happening, the Beijing government uses *hukou* to control and divide people into locals and migrants. People entitled to get the Beijing *hukou* are the local people and thus, they could enjoy these social resources. This is very much a national situation in China.

In Beijing, the majority of migrants rent housing in the neighborhood. Considering that the migrants' financial status is not very good, usually the migrants rent houses which impose low rental. In Beijing, a lot of neighborhoods accommodate more than 50% rural migrant residents (He, 2008).

Also, a few migrants live in the housing provided by their companies. Usually these houses are provided for laborers working on a public project, but with bad housing facilities. These people constitute 1.0% ~ 2.0% of the total number of migrants. The houses are very crowded and facilities are poor, where being without a bathroom and kitchen is very commonplace.

Overall, the migrants in Beijing lack the basic housing rights and cannot enjoy any housing policy that they may well deserve. The best choice for them is to rent low – rented houses in the neighborhood. These houses suffer from housing overcrowding, privacy affected and poor housing facilities. Therefore, the housing system has been unfair for migrants in Beijing simply because they do not gain any access to the Beijing *hukou*.

3.5 Summary

This chapter mainly focuses on the improvement of housing system in China from 1949 until now. With the rapid economic development, the housing system becomes more and more useful to adapt to the situation of China. With the help of housing system development, the housing characteristics (Overcrowding, housing privacy and housing facility) situations improve a lot in urban area. However, the only drawbacks of the housing system only focus on the local residents who hold the local ID and without consider the other migrants. Therefore, the researcher hope this drawback should be avoided and to treat everyone (local and migrants) fairly. Thus, the migrants could enjoy the housing in urban area and their housing characteristics and living condition could be improved.

Chapter 4

Research Objectives and Methodology

4.1 Introduction

This chapter discusses the overall approach taken by the researcher and the development of the objectives. Primary data were collected through a survey on rural migrants in Beijing Fengtai District using questionnaires while secondary data was gathered from utilizing the written materials (including some published documents), file material, reports and so on. Univariate and bivariate analysis methods were used to examine the primary data and the results were discussed using the descriptive and inferential statistics like comparing the means (Independent T Test), correlations and regression etc.

As in previous empirical studies on this subject, the research on the relationship between housing characteristics and living condition among rural migrants has to be both descriptive and exploratory not only because of the lack of adequate data and information but also because of the limited amount of systematic research on the subject undertaken in China.

4.2 Research Approach/Process

4.2.1 Approach

Through literature review, the researcher finds that very few researchers have done these researches which are related to housing characteristics and living condition, so it is necessary to approach this study in the overall context of urban activities. Examination of the past and present rural migration in China is crucial in order to paint an appropriate picture so as to establish the extent of knowing the rural migrants' housing characteristics and their effects on living condition.

While the paucity of existing material determines the overall methodology of the empirical study, the specific problem addressed necessitates the use of questionnaires and interview. So the quantitative and qualitative approaches will be used in this research. For the use of questionnaires (quantitative approach), the researcher had chosen a selective sample among the rural migrants in Beijing Fengtai District. A questionnaire, to be filled in by the respondents was considered as the most appropriate tool for gathering data from rural migrants since it would provide uniformly collected data. For the use of interview (qualitative approach), the researcher could get the information about rural migrants and their housing data from government official and unofficial publications. The quantitative (questionnaire) and qualitative (interview) approaches were viewed as suitable for this study because of the descriptive nature of the subject area.

4.2.2 Process

A research on the influence of housing characteristics on rural migrants' living condition seemed appropriate since its significant impact on the urbanization activity has received little academic attention all over the world. According to previous studies completed in various countries both by the government and researchers, although self-investigation has formed a major portion of these activities, it was a neglected research because of the absence of detailed and systematic studies.

The theoretical basis for this research relates to the rural migrants and housing characteristics as well as their living condition. It is the intention of this study to examine the influence of housing characteristics on rural migrants' living condition. For this purpose, it is essential to examine variables that are related to the overall conceptual framework of this activity.

This research employed a survey method to collect the data pertaining to rural migrants and their housing characteristics as well as living condition. The population of rural migrants totals around 353 009 (BCD, 2012) for the survey it is suggested that 400 samples were to be used, as according to the figure by Robert & Daryle (1970) it suggests that the population of 75 000 needs sample size of 382 and population of 1 000 000 needs sample size of 384. The survey was completed in four parts of Beijing Fengtai District which is divided by the Beijing Ring Roads. According to the "Beijing Construction Department" (2012), the population of the rural migrants in each part is around 88 000. Since the study concentrates on Beijing Fengtai District, each part will find 100 rural migrants as the representatives of this population. Due to the rural migrants prefer to live in clusters in the neighborhood, so it is easily to identify them to do the research.

The primary data collected were processed using a statistical package to acquire results which were analyzed by applying the univariate and bivariate analyses (refer to Figs. 4.1~4.2).

4.3 Research Objectives and Research Questions

4.3.1 Research Objectives

(1) To identify the rural migrants, type of housing, housing characteristics and living condition;

(2) To examine the migrant housing characteristics in Beijing Fengtai District;

(3) To examine the effects of housing characteristics on rural migrants' living condition in Beijing Fengtai District.

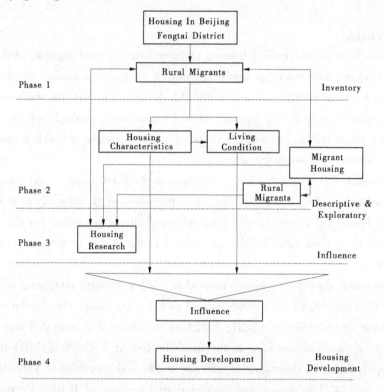

Fig. 4.1 Detail research process flowchart of work plan I

4.3.2 Research Questions

(1) What is migrant housing? Where are their locations? What types of housing?

(2) Who are the rural migrants? What are their profiles? What are their occupational history?

(3) What are the housing characteristics in Beijing Fengtai District? What are their living condition?

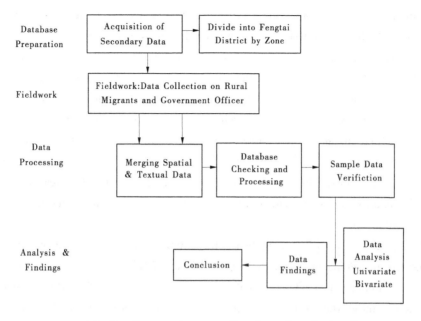

Fig. 4.2 Detail research process flowchart of work plan Ⅱ

(4) How do housing characteristics affect the rural migrants' living condition in Beijing Fengtai District?

4.4 Definition

4.4.1 Rural Migrants

Zhang and Hou (2009) provide a concise comparative definition of rural migrants. They state that rural migrants have some notable characteristics: the population number increases every year; the majority of them perform small businesses in urban areas; a large majority of rural migrants' age is below 30; the education level is very low etc. Also, they emphasized that the big cities, especially the metropolis cities like Beijing, Shanghai, Guangzhou and Shenzhen will become the gathering area for rural migrants. Until 2000, the rural migrants in China reached to 97 000 000, such a big number of rural migrants influence the economy, society and housing in urban areas (CCG, 2000). Until 2009, two thirds of the population is rural migrants in Beijing and Shanghai. In the southern part of China, some small cities could even reach 1 : 1 between the local population and rural migrants. Zhang and Hou (2009) notice that these rural migrants want to escape from rural areas and make their way into urban areas to try to improve their lives.

Wu (2010) has defined rural migrants as workers or small businessmen; who migrated to urban areas with the hope of finding better economic opportunities. He also found that in urban areas, 95% of the rural migrants rent and live in the neighborhood. In China, the neighborhood usually refers to many housing units inside one neighborhood together with independent facilities. Residents live in the neighborhood could interacted with each other outside their houses. In China, it is reported that rural migrants in urban areas usually rent the houses in neighborhood with low rental. Due to the rural migrants without Beijing *hukou*, therefore, they need to rent house in the neighborhood among local residents. Hong Kong authorities pay much more attention on rural migrants by including their living places, living condition and housing characteristics, also they state whether they are urban or rural migrants, when they migrate to Hong Kong and when they leave. Therefore, the rural migrants' housing characteristics and living condition in China also should be given due attention.

Beijing Migrant Registration System defines rural migrants as who migrate from rural area and live in Beijing more than 1 month (BMRS, 2010). Also, as mentioned by the previous researcher Wan (2005), it states that rural migrants are divided into 2 categories which include those who are temporary rural migrants and permanent rural migrants. Temporary rural migrants refer to rural migrants staying in urban areas for more than 1 month but less than 6 months and they will finally return to their hometown. Meanwhile, permanent rural migrants refer to rural migrants staying in urban areas for a long time, usually more than 1 year. In this research, the researcher will choose temporary rural migrants and permanent rural migrants both. Thus, the rural migrants who live in Beijing Fengtai District more than 1 month are eligible to be the respondents. For whether they will return to their hometown or not, it will not affect the direction of this research.

4.4.2 Housing Characteristics

In "US Housing Characteristics Census: 2000" that:

"A housing unit is a house, an apartment, a mobile home, a group of rooms, or a single room occupied, or intended for occupancy, as separate living quarters. Separate living quarters are those in which the occupant(s) live separately from any other people in the building and which have direct access from outside the building or through a common hall."

The "US Housing Characteristics Census: 2000" further mention that human's living condition and living quality are all related to housing characteristics such as housing overcrowding characteristic, privacy and facility etc (USDHUD, 2007). Human's quality of life is affected by the housing characteristics. Good housing characteristics will contribute for better living condition. In Beijing, a city with large number of rural migrants' population, the housing characteristics should be made the focal point. Among thousands of housing characteristics that can be listed down, the housing overcrowding characteristic, housing privacy characteristic and housing facility characteristic would constitute serious issues for migrants, especially rural migrants (Jiang, 2006; Wu, 2002; Jian & Ye, 2003).

Of course, there are many other housing characteristics in the developing and developed countries both. In some developed countries like Germany, UK, Japan, the housing characteristics mostly focus on interior air quality, air ventilation, housing design, housing technology and green building etc. With the supportive of strong economy and high technology, the housing characteristics could be easily updated. While, due to the low economy together with large population, the rural migrants in developing countries only focus on the basic housing characteristics like overcrowding, privacy and facility. Low economy and high expenditure will restrict them to pursue the better housing characteristics (Pang, 2003).

Evidently, to successfully solve these housing characteristics problems their housing characteristics and the situations in which they live need to be improved. In turn, such moves will contribute to the betterment of their lives as a whole.

4.4.3 Living Condition

Living condition refers to the factors that could influence human's life either inside or outside the house (Jiang, 2006). Housing characteristics is one of the factors placed under living condition and as they could influence living condition. According to Havel (1957), there are two problems that emerge between the living condition and housing:

(1) Human needs to be satisfied about their houses;

(2) Housing attempts to satisfy human needs, but they happen o be restricted by technology, economy and law.

Sometimes people are always confused with "living condition" and "housing condition" as the two terminologies, but living condition contains much more meaning than housing condition. Havel (1957) asserts that living condition is for people to perform various activities, work, entertain, meal, rest or sleep. France Institute defines living condition as "the geographical circumstance of which suits human, animal and plant so that they can survive there"; the definition indicates that it also has an ecological meaning. Overall, good living condition is habitability that could provide human with good sleep, work, relax and good environment inside or around the houses. Also, the good and comprehensive social facility also needed to provide with good living condition.

So what is a good living condition that could suit human culture and civilization? A good housing constructed in 1950 in France had had good decoration, good ventilation, good day lighting, although without bathroom. In that period, that was already a very good house but now people cannot bear being in a house without a bathroom. Now human considers housing with a bathroom, kitchen, living room and bedroom which are the basic necessity. But with the advancement of housing development, human's housing requirements also increase. Now human wish housing should have balcony, warehouse, garage and closet etc. From the above discussion could know that living condition standard increase together with social and housing development, it could be regarded as the representation of human's civilization.

Overall, humans should have good living condition whether inside or around the houses. Good living condition should well equip with good housing facilities and the social facilities around the houses. Also, good living condition should have good housing characteristics. To improve human's living requirements, housing characteristics and living condition both need to improve.

4.5 Data and Information Collection

4.5.1 Primary Source

Primary data were obtained using a questionnaire which was completed by interviewers. The interviews were completed by the researcher with the help of six other government officers (Beijing Construction Department, Beijing Migration Control System and Beijing Migrant Household Registration System). Observation was also made to provide information which was not covered in the questionnaire such as the overall selected sites' cleanliness, housing overcrowding and facility situation. With the help of observation, the researcher could easily know the housing environment around their houses. It includes the sanitation problem, social facility situation and even the social security problem. Through observation, the researcher found that the sanitation situation is quite unsatisfied. Also, many social facility is very far from the houses or do not have. By the way, some rural migrants mentioned that their bicycle always be stolen. It represents that the social security have some problem.

The questionnaire was distributed to the respondents and filled in their own housing units during the fieldwork. A total of 400 questionnaires were completed which is a 100% return rate.

a. Population

The population of this study was made up of 40 neighborhoods in four parts of Beijing Fengtai District where 30% of residents in each neighborhood are rural migrants. In summary, the sampling procedure applied is shown by Table 4.1 below.

Chapter 4　　　　　　　　　　Research Objectives and Methodology

Table 4.1　Summary of sampling procedures

Four research areas in Beijing fengtai district	Population number of rural migrant	Number of neighborhood in each research area	Number of neighborhood (refer to more than 30% of the population are rural migrants)	The number of total sample survey neighborhood
Research area 1 (between second and third Beijing ring road)	Estimate 90 000	31	12	10
Research area 2 (between third and fourth Beijing ring road)	Estimate 87 000	27	11	10
Research area 3 (between fourth and fifth Beijing ring road)	Estimate 88 000	25	13	10
Research area 4 (outside fifth Beijing ring road)	Estimate 87 000	20	13	10
Total	Estimate 352 000	103	49	40

Source: 1.*Beijing Construction Department*, 2012.
　　　　2.*Beijing Migration Control System*, 2012.

b.Sampling frame

A list of neighborhoods with a number of rural migrants to be used as the sampling frame for the study was obtained from the Beijing Construction Department and Beijing Migration Control System of both councils of Beijing. Documents containing this information were obtained from the Beijing Fengtai District map which was further categorized into four research areas(see Fig. 4.3).

Beijing Fengtai District is divided into four research areas by Beijing Ring Road (refer to Table 4.1). Since the rural migrants reside everywhere in the four parts of Fengtai District, it is important for the samples to cover these areas. Moreover, the periodic nature of rural migrants requires the study to cover the area in which these rural migrants were located as much as possible. Appendix 1 shows the locations of 40 neighborhoods in Beijng Fengtai District, also each research area contains almost the same number of the neighborhood with more than 30% residents are rural migrants in each neighborhood. For example, research area 1 contains 12 neighborhoods, the research area 2 contains 11 neighbor-

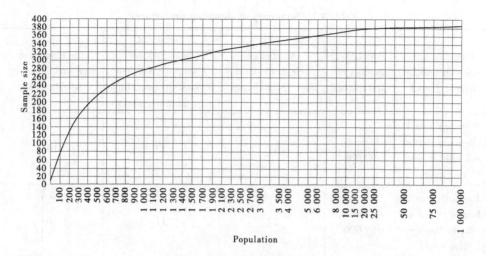

Fig.4.3 Determining sample size
Source: Robert V. Krejcie and Daryle W. Morgan(1970). Determining sample size for research activities.

hoods, the research area 3 contains 13 neighborhoods, and research area 4 contains 13 neighborhoods. Therefore, the researcher chooses 10 neighborhoods in each research area. Totally 40 neighborhoods will be chosen in this research. Details of the distribution of neighborhoods in each research area are shown in Appendix 2, which was used as the basis for the sampling frame of the sites selected.

c.Sample size

In order to obtain proper representatives of the said locations, Beijing Fengtai District was divided into four parts: Research Area 1; Research Area 2; Research Area 3; Research Area 4. According to the Beijing Construction Department and Beijing Migration Control System, a list of neighborhoods which covered the four parts of research areas in Beijing Fengtai District was formed. 40 neighborhoods (more than 30% of population are rural migrants) were chosen as the target.

The selected neighborhoods also show the housing number and population number in each selected neighborhood that easily helps researcher to calculate the samples. Appendix 2 shows the selected neighborhoods in each research area with a total number of samples. According to the sampling methods suggested by Krejcie and Morgan (1970), 400 research samples were chosen. Also, due to the rural migrant population number in each research area almost same (around 88 000). Therefore, the researcher chooses 100 respondents in each research area.

Financial and time costs were major constraints in deciding the sample size, bearing in mind the cost of traveling from one neighborhood to another meeting and enquiring the neighborhood manager.

d. Duration of survey

The surveys were completed in 3 months to cover all selected locations. However, because of rural migrants' low education and non-cooperation nature which needed some slack time, the survey took longer than expected.

e. Pilot survey

A pilot survey was conducted to access the quality of the questionnaire. According to Hertzog (2008), the pilot survey should use more than 15.0 % of the total samples to increase its accuracy. For this purpose, the researcher had chosen 20 respondents in each research area to a total of 80 respondents (20.0% × 400) in Beijing Fengtai District. On average, the researcher asked 2 respondents in each selected neighborhood. Due to longer traveling distances from one neighborhood to another, the pilot survey took 40 days to complete. After obtaining the data, the researcher used SPSS (Statistic Package of Service Solution) software to perform a reliable test- usually the Cronbach's Alpha of more than 0.7 to represent data with high reliability. In this research, the Cronbach's Alpha is 0.833, more than 0.7 and it represents that the questionnaire is highly reliable (see Table 4.2).

Table 4.2 **Reliability statistics (SPSS output)**

Cronbach's alpha	Cronbach's alpha based on standardized items	N of items
0.833	0.838	71

The pilot survey was valuable in order to clarify the appropriate questionnaire format. Experiences from previous studies show that personal or demographic questions should be posed towards the end of the questionnaire (Babbie 1983). Nevertheless, Babbie (1983) has found out that the respondents had found it tedious to answer the open-ended questions especially if people are very busy, so the open ended questions will be shortened. In the event of such questions about their family backgrounds were considered as conversation openers and these were asked at the beginning when rural migrants seemed to be more relaxed. Another important function of the pilot survey was to determine the best time of the day to conduct the survey. It was found that the best time was around 12 p.m., just after respondents had finished working and returned to their houses. Furthermore, it was also their least busy time of the day while at other times they were always working. The pilot survey was completed 40 days before the survey format was finalized.

4.5.2 Secondary Data

Through secondary data, the researcher will define the rural migrants and the type of housing in which they live first, especially *where* they live and how they could be identified. Then, the researcher will understand the overall situation of the housing characteristics and living condition for migrants all over the world, especially in China. Then, through the secondary data the researcher will form the questionnaire and interview questions to determine the real situation behind the rural migrants' housing characteristics and the effects on their living condition.

The collection of secondary data was made possible as references were made from published articles, journals, reports, books and magazine. Besides, some information or data can be obtained from the Internet.

Information pertaining to the studies that was not covered in the questionnaire and interview questions was obtained from secondary sources. Information concerning housing characteristics and the living condition of the rural migrants was procured from various research publications and some government publications. Some secondary sources from government included:

(1) Beijing Construction Department (2005, 2006, 2010, 2012)
(2) China and Beijing Housing Policy (2012)
(3) Beijing Migrant Household Registration System (*hukou* System) (2010, 2012)
(4) Beijing Migration Control System (2012)
(5) Beijing Government (2012)

4.6 Data and Information Analysis

The analysis of this study was both descriptive and explanatory. In describing the housing characteristics and the effects on living condition, descriptive statistical techniques, such as means, chi-square and correlations were used. Also, inferential statistics like the comparison of means, bivariate analysis and regression were used in the data analysis to measure some dependent variables are related or not.

The second section of the analysis dealt with the key issues at hand as a basis for assessing whether housing characteristics or not could influence living condition. The cross-tabulation between variables to establish the strength and weaknesses of relationships was one of the tools used in the bivariate analysis. For the non-parametric analysis, a chi-square test of significance will be used to determine whether or not a set of two variables were related. Generally, a chi-square value indicates that the test is significant and the variables are not independent; i.e. the variables are related. The strength of the relationship can be determined using chi-square based measures or proportional-reduction-in error measures of associations. Since the 2×2 tables were used for the cross-tabulation of the variables, a Pearson chi-square was employed. The probability (observed significance level) should be small enough (normally less than 0.05 or 0.1) to reject that the variables are independent.

4.7 Research Constraints

4.7.1 Limitations of the Study

This study was confined to the study of the rural migrants in Beijing Fengtai District with the analysis based on data collected only from this area. Although the results may be applicable to other rural migrants elsewhere, it is not suggested that the sample used speaks for rural migrants elsewhere in China or in other countries. However, the implications of the findings are fairly universal in that any proposal to develop similar type of research elsewhere could no doubt benefit from the findings obtained from this investigation.

The research was restricted to the analysis of the rural migrants' locations in Beijing Fengtai District which were registered with the government department. Since it was difficult to obtain the name of the neighborhood in which the majority of the residents are rural migrants, it was initially assumed that all rural migrants were registered with the government department. However, later on, the researcher has managed to get his data through interview with the government officers.

The paucity of the written material and documentation about rural migrants has also posed some constraints to the study. The failure of the government departments including both the Beijing Construction Department and the Beijing Migration Control System to provide substantial secondary information had required the researcher to rely mainly on primary data. Furthermore, the fact that Wu (2010) completed work on the floating population in China was the latest analytical work available has also indicated the importance of collecting new primary data for this study.

4.7.2 Fieldwork Problems

The first problem concerns with the unreliability of the information from the government departments. The sampling of the neighborhoods was based on the rural migrants provided by the government departments. However, since the data were not updated prior to the survey, some of the neighborhoods had already changed names without the knowledge of the government departments. As this became known only in the neighborhoods, the researcher had to find the locations by himself. Such incidents had indirectly led to the extensive use of valuable time and resources.

Another problem dwells into the non co-operation of some respondents. In the neighborhoods, some rural migrants had refused to be interviewed. Some rural migrants would give an excuse of I am busy or Cannot interview me. As a result, the researcher had to move on to another rural migrant. Usually some pens or pencils had been given to the respondents as tokens of appreciation. Although the number of such non-respondents was low, their non-co-operative attitudes did affect the survey process to some extent since other rural migrants became suspicious of the purposes intended by the researcher.

Because Beijing Fengtai District is very big and the survey was carried out in summer, the great distance from one neighborhood to another and hot weather posed some difficulties to the researcher. Especially during the working days (Monday ~ Friday), the serious traffic jam had been one of the daunting challenges as well.

Fortunately, these problems did not affect the overall findings of the study since most were rectified during the fieldwork.

4.8 Questionnaire Design

4.8.1 Rural Migrants

The design of the questionnaire was guided by the research problem and survey objective. Since the objective of the survey was to have the interviews conducted in a uniform way, formal or structured interviews were employed. In order to avoid bias resulting from the questionnaire design, the questions were constructed in such a way that they were direct, simple and familiar to the respondents. However, some explanations by the interviewers were necessary to clarify certain points so that a certain level of consistency could be achieved. The questionnaire was divided into four sections including:

Part A. Respondents' Personal Data
Part B. Occupational History
Part C. Housing Characteristics
Part D. Influence on Living Condition

The details of the questionnaire are shown in Appendix 3. To measure the personal data, it could help the researcher to define the rural migrants (gender, age, education etc.) and to relate these variables to each other. Also, the occupation history could easily reveal the economic situation of the rural migrants in Beijing and relate to their personal data, the bivariate analysis could be clearer to achieve the stated objectives. The housing characteristics and influences on living condition use correlation, regression and T-test to measure these variables. If it is proven to be statistically significance then the cross-tabulations are utilized.

4.8.2 The Government Department

The semi-structured questions were targeted at the government department officers in Beijing Fengtai District who monitor the rural migrants and the housing that they live; they are the Beijing Construction Department, Beijing Migrant Registration System and Beijing Migration Control System. The questions in the interview, which were based on the migration guidelines produced by the government, were aimed at obtaining information about the rural migrants. They were related to the registration process of the rural migrants, the housing characteristics and living condition of these migrants.

The details of the semi-structured questionnaire are as shown in Appendix 4. The semi-structured interview could help the researcher to know the government's methods of approval, registration and management in tackling the housing problems of rural migrants in Beijing. Also, it could help the researcher in obtaining secondary data on locations and distributions of the migrants, population as well as providing the sampling frame for the survey. The semi-structured questions provide great assistances for the researcher to carry out this research.

4.9 Summary

In this chapter, the research approach, research process, objective and research questions were provided. Also, the definition of rural migrant, housing characteristic and living condition were given to make a clear picture about the research. Detail data collection and data analysis could help the researcher to carry out this research easily. In the research limitation and field work problem, the researcher mentioned the difficulties of this research and future research willingness. The questionnaire for migrants and the semi-structured questions were also elaborated in this chapter and guide the researches to carry out the research.

Chapter 5
Analysis and Results

5.1 Introduction

In many cities in the Asian region, rural migrants are commonly portrayed as unskilled, poorly educated people surviving in the big cities. Rural migrants also search for a temporary measure of survival whilst awaiting better opportunities in big cities. In China, there are approximately ninety-seven million rural migrants up until now (CCG, 2000).

Some perceive rural migrants as ones who hold a temporary occupation and survive in the urban areas. Their housing characteristics are very poor and their living condition is deteriorated. Asami (2005), in describing various attitudes towards rural migrants, has pointed out that at the extreme, rural migrants are portrayed as shrewd businessmen who try to get more financial resources in urban areas for their future life. Housing characteristics and living conditions are critical points to determine their lives in the city. It therefore appears necessary to know these points using the analysis of the demographic data. Such an analysis is also necessary to associate the profiles and characteristics of their housing and living condition.

It has been shown that the relationship between housing characteristics and living condition is considered to be one of the most important linkages associated with rural migrants. It is assumed that housing characteristics could affect rural migrants' living condition. In addition, improvement methods on the housing system in Beijing could play a significant role in improving rural migrants' housing characteristics and living condition. Government policy should be fair to everybody, and this includes rural and urban migrants.

Basically, a general question to be asked is whether it is worthwhile for the rural migrants to leave their hometowns and migrate to big cities. Could the rural migrants find suitable jobs in big cities? Could the current housing characteristics influence rural migrants' living condition? At what point will their housing need to improve and benefit everyone?

5.2 Research Area

5.2.1 Fengtai District

The research area was located in Beijing Fengtai District, a suburban district of the municipality of Beijing. It lies to the southwest of the urban core of the city. Fengtai District is located in the south-west of Beijing and is the second largest district in Beijing.

The southwestern part stretches into the 3rd Ring Road, 4th Ring Road and 5th Ring Road (Ring Road is a circle around Beijing which is seen as a solution to Beijing's traffic problems) running through the area. It is 306 square kilometers in the area and is home to1 360 000 inhabitants (BTG, 2011). Until 1990, during the rapid economic development, Beijing Fengtai District starts to urbanized and becomes the urban center in Beijing (see Fig.5.1).

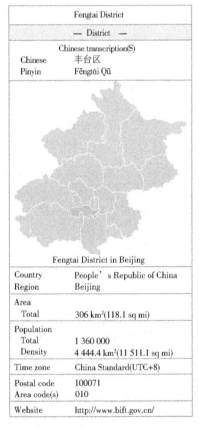

Fig. 5.1 Fengtai District location map
Source: Beijing Travel Guide (2011).

The Housing Condition of Rural Migrant in Bejing Fengtai District

Until 2011, the whole population in Beijing Fengtai District totaled 1 360 000, while the population of migrants in Fengtai District reached 489 000 (BTG, 2011). The number of rural migrants is 72.19% (BCD, 2012) among the total migrants of 489 000 in Beijing Fengtai District. Such a large amount of rural migrants live and work in Beijing Fengtai District without Beijing *hukou* (household registration system) (Zhu, 2007; Jiang, 2006; Xu, 2005; Zhang, 1995), It has been reported that there are a lot of living condition problems for this special group, including the fact that their settlements are very crowded, that they have no privacy and facilities incomplete (Jiang, 2006; Boyd, 1989; Goldscheider, 1983; Lipset and Bendix, 2001) (see Fig. 5.2).

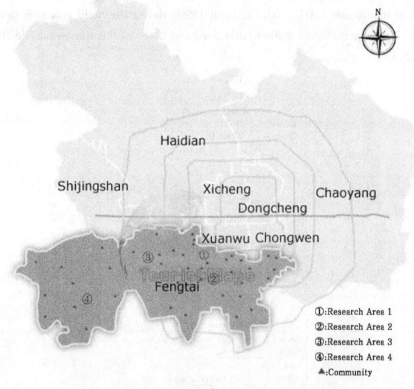

Fig. 5.2　The map of Beijing urban area
Source: Google Map.

5.2.2　Neighborhood

This research is conducted on 40 neighborhoods in Beijing Fengtai District. Due to the fact that the Fengtai District was divided by Beijing Ring Road into four parts, this research will then be carried out in the four areas separately (see Table 5.1).

Chapter 5 Analysis and Results

Table 5.1 Summary of sampling procedures

Four research areas in Beijing fengtai district	Population number of rural migrant	Number of neighborhood in each research area	Number of neighborhood (refer to more than 30% of the population are rural migrants)	The number of total sample survey neighborhood
Research area 1 (between second and third Beijing ring road)	Estimate 90 000	31	12	10
Research area 2 (between third and fourth Beijing ring road)	Estimate 87 000	27	11	10
Research area 3 (between fourth and fifth Beijing ring road)	Estimate 88 000	25	13	10
Research area 4 (outside fifth Beijing ring road)	Estimate 87 000	20	13	10
Total	Estimate 352 000	103	49	40

Source: 1. *Beijing Construction Department*, 2012.
2. *Beijing Migration Control System*, 2012.

The detailed location of the neighborhoods in Beijing Fengtai District will be shown using the Google map, 2012 (Fig. 5.3 until Fig. 5.6).

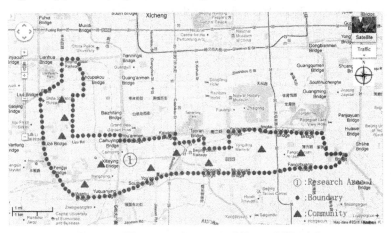

Fig. 5.3 The research area 1 of Fengtai District
Source: Google Map and Beijing Migration Control System, 2012.

· 83 ·

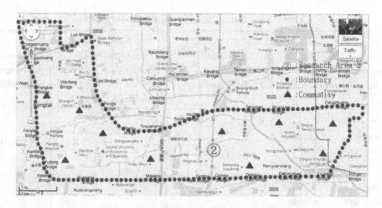

Fig. 5.4 The research area 2 of Fengtai District
Source: Google Map and Beijing Migration Control System, 2012.

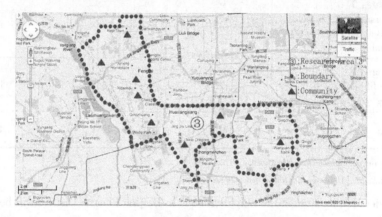

Fig. 5.5 The research area 3 of Fengtai District
Source: Google Map and Beijing Migration Control System, 2012.

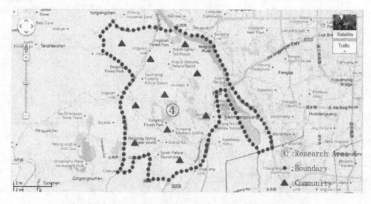

Fig. 5.6 The research area 4 of Fengtai District
Source: Google Map and Beijing Migration Control System, 2012.

5.2.3 Housing

As most rural migrants are without Beijing *hukou* (local ID), so the best choice is for them to rent houses available in the neighborhood. In Beijing, 90% of rural migrants choose to rent houses in the neighborhood (Sanders, 1986; Zhang, 2007; Zheng, 2004; Zhong, 2000; Zhou, 2003). Linked with the fact that the rental is often cheap, these communities are typically old, dirty and the facilities abysmal. (see Plate 5.1 until Plate 5.3)

Plate 5.1 Fang Qun Yuan community
Source: Camera by Researcher.

During the survey, the researcher notices that these neighborhoods were built in the 1960's or 1970's. These neighborhoods are middle high-rise buildings. The floor is between 4~6. More than three decades of living, together without the care-taker, have contributed to why these neighborhoods become old and shabby. The housing facilities are out of fashion and they can no longer serve the residents in these communities. Poor design is translated into dark lighted and overcrowded rooms.

5.3 Rural Migrant

5.3.1 Personal Data

In this research, it is important to know the rural migrants' personal data like gender, marital status, age, income status, education level and geographic distributions etc. And use Chi-Square Test, correlation analysis, compare mean (T-Test) and regression analysis to determine the statistically significance of these variables. Then cross-tabulation is used to represent the related variables. From these variables, the researcher could easily identify the profile of the respondents. It is therefore significant to do the analysis and discussion on these variables to be related to other variables.

Plate 5.2　**Xing He Yuan community**
Source：Camera by Researcher.

Plate 5.3　**Xin Xing Jia Yuan community**
Source：Camera by Researcher.

5.3.1.1 Age Distribution

Through Pearson Chi-Square Test, the Sig. 0.016 less than 0.05 shows that age group and marital status has statistically significance and could use the cross-tabulation to discuss it. As shown in Table 5.2, the majority of the rural migrants are less than 42 years old, which is regarded as the most dominant age group. These rural migrants migrate to Beijing to search for job opportunities and live here for several years. In this age period, there are many children, young and middle age rural migrants. They come to Beijing to pursue the occupation or do some small business. As previous research (Harris, 1991; Zhou, 2002; Zhu, 2001; Zhu, 2002; Zong, 2007) suggests, the rural migrants prefer to do migration in their young and middle age and strive to compete for their future life. At the same time, many rural migrants mentioned in the survey that they want to earn more money and "open their eyes" (enhance life experiences) when they are young.

Furthermore, the age group lies in the 31~42 years old category, which is regarded as the rural migrants' most productive age group. At this age, the rural migrants are not only expected to be the primary earners of the household but also to be more or less at a settled stage in their lives. It is not surprising to find that the majority of the rural migrants within this group were already married (see Table 5.2).

Table 5.2 **Age group and marital status of the rural migrants**

Age of rural migrants	Single	Married	Divorce	Total
Less than 18 years	136	7	0	143
%	95.1	4.9	0	100
More than 18~30 years	30	56	9	95
%	31.6	59.0	9.4	100
More than 31~42 years	25	79	15	119
%	21.0	66.4	12.6	100
More than 43~54 years	3	21	4	28
%	11.0	75.0	14.0	100
More than 54 years	1	12	2	15
%	7.0	80.0	13.0	100
Total	195	175	30	400
%	48.5	44.0	7.5	100

(Pearson Chi-Square Test: Sig. 0.016 < 0.05)

However, more than one third of the rural migrants are less than 18 years of age and it represented the popularity and necessity of migration to the young in China.. According to Zhu (2007) that more than half of the children (less than 18) quit schooling in most rural areas of China. Migrating to other big cities such as Beijing deems to be an alternate choice for them to move on in life. Nonetheless, without proper education they find it difficult to find good paying jobs to support their life, and as a result, their future development may be affected.

The age group lies in the 18~30 years old category, which represents that many of the migrants are young people. These people migrate to urban areas searching for a better opportunity to improve their lives. During the survey, some rural migrants in this age group mentioned some of them were divorced due to lack of stable income.

5.3.1.2 Gender Ratio and Income Status

In their study of rural migrants in Beijing and Shanghai, Wang (2004) and Wan (2005) find that there are more male than female rural migrants in big cities. Similarly, Liu (2005) in his discussion on the urban migration points out that "male rural migrants are more likely to migrate". This study in Beijing Fengtai District, shows that the rural to urban migration was also male dominated (see Table 5.3).

Table 5.3 Male and female composition of rural migrants

Gender	Number of rural migrants	%
Male	260	65.0
Female	140	35.0
Total	400	100

One of the reasons for the male dominance in urban migration appears to be related to the nature of the male himself. The male has the ability to support their families and migrate to urban areas. Usually, male rural migrants would leave their families behind and migrate alone. As some rural migrants had mentioned, they migrated to Beijing to earn money. Finally, they would bring the money back to their hometown. However, some rural migrants had brought their whole family to Beijing. Their wives still stay at home and take care of the children. However, these findings do not nullify women's involvement in working as they are very likely to be involved actively as supporters of the family. A typical example was, female rural migrants still are working and earning as much as their male counterparts. As shown in Table 5.4, the most dominant income group lies within the

1 001 ~ 3 000 RMB per month. However, the average income in Beijing was 5 223 RMB per month in 2013(BMBS, 2013). Therefore, it is very difficult to support their families in Beijing. The second dominant income group lies in the < 1 000 RMB per month. So, it is very difficult for them to survive in Beijing for house and food. Only 12.0% of rural migrants' income is between 5 001 ~ 8 000 RMB. Some of them have mentioned that they could save some money and send it back to their hometowns. 3.5% of the rural migrants' income total more than 8,000 and this figure includes three women. Therefore, the female rural migrants' working ability should not be underestimated.

Table 5.4 Income status of male and female rural migrants (RMB)

Income status	Male	Female	Total
< 1 000	65	39	104
%	62.5	37.5	100
1 001 ~ 3 000	98	51	149
%	65.8	34.2	100
3 001 ~ 5 000	55	30	85
%	64.7	35.3	100
5 001 ~ 8 000	31	17	48
%	64.6	35.4	100
> 8 001	11	3	14
%	78.6	21.4	100
Total	260	140	400
%	65.0	35.0	100

(Pearson Chi-Square Test: Sig. 0.037 < 0.05)

Overall, the male respondents' average income is 2 900 RMB and female respondents' average income is 2 300 RMB. And the total respondents' average income is 2 600 RMB. It represents that the rural migrants' average income is much lower than the average income in Beijing 5 223 RMB per month in 2013(BMBS, 2013). So, it is very difficult to support their life in Beijing.

5.3.1.3 Gender and Education Level

Through Pearson Chi-Square Test, the Sig. 0.000 less than 0.05 show that gender and education level has statistically significance and could use the cross-tabulation to discuss it. A number of previous urban migration studies (Li, 2004; Li, 2006; Davis, 1990) have shown that education levels were generally low and that the majority of rural migrants had no formal schooling. Also, studies completed by Cai (2002) and Ding (2001) in Shanghai, have both shown that the majority of the rural migrants had at least only completed their primary education. Although these studies were completed in other areas, the findings are comparable to this study which is shown in Table 5.5.

Table 5.5 Gender and education level of the rural migrants

Education level	Male	Female	Total
No formal education	85	20	105
%	81.0	19.0	100
Primary school	60	40	100
%	60.0	40.0	100
Secondary school	55	45	100
%	55.0	45.0	100
Diploma	20	20	40
%	50.0	50.0	100
Degree	25	5	30
%	83.3	16.7	100
Postgraduate	15	10	25
%	60.0	40.0	100
Total	260	140	400
%	65.0	35.0	100

(Pearson Chi-Square Test: Sig. 0.000 < 0.05)

As expected this shows that the rural migrants' education level in Beijing is low and only holds primary or secondary education level, or even without schooling. But around 24% (95 respondents) have post school levels-diploma to postgraduate, it represented that the rural migrants' education level is not low compare to the previous research. So, this research came out with some different result.

Also, the education level is highly related to the rural migrants' application for a Beijing ID (*hukou*). The Pearson Chi-Square Test, the Sig. 0.012 less than 0.05 show that education level and *hukou* has statistically significance. From this research, the majority (76.0%) of the rural migrants who obtain the low education level are quite difficult to get the Beijing ID (*hukou*). However, with the education level increases, the rural migrants could get the Beijing ID (*hukou*) easily. Not surprisingly, the rural migrants who hold postgraduate degree could get *hukou* very easily and 84.0% of them did. In the survey, the researcher noticed that these "more educated" rural migrants live in Beijing for more than 5 years and work in the government department or big company. Therefore, higher education levels could help the rural migrants to find better jobs and obtain the Beijing ID (*hukou*) easily (see Table 5.6).

Table 5.6 Education level and Beijing ID (*hukou*) status of the rural migrants

Education level	Get *hukou*	Without *hukou*
No formal education	0	105
Primary school	0	100
Secondary school	2	98
Diploma	5	35
Degree	12	18
Postgraduate	21	4
Total	40	360
%	10.0%	90.0%

(Pearson Chi-Square Test: Sig. 0.012 < 0.05)

5.3.1.4 Length of Stay in Beijing

As Table 5.7 shows, the vast majority of rural migrants have lived in Beijing for less than 1 year. During the survey, these rural migrants all mentioned that they migrate to Beijing due to Beijing's distances to their hometown. But due to the dry weather and housing overcrowding in Beijing, they feel they are not suited to work here and want to migrate to other areas like Shanghai or Guangzhou. Therefore, these people want to migrate to the southern area of China later. So, more than one third (34.0%) of the rural migrants only stay in Beijing less than 1 year and migrate to other area due to the weather and overcrowding problem. So, Beijing as a transit place before they can move elsewhere.

Table 5.7 When the rural migrants migrate to Fengtai District (%)

Year	Frequency	Percentage
Less than 1 year	136	34.0
1 ~ 2 years	101	25.3
2 ~ 5 years	80	20.0
5 ~ 10 years	40	10.0
10 years before	43	10.8
Total	400	100.0
Average		3.5 years

However, 25.3% of the rural migrants have stayed in Beijing for 1 ~ 2 years. They mentioned that the main reason for migrating to Beijing was to earn a living and to support their other families in their hometowns. Thus, these rural migrants' migration period in Beijing is only on a temporary basis. However about 21% of the migrants have stayed in Beijing for longer periods, which is between 5 to more than 10 years. They mentioned that, the satisfactions of their job situations have made Beijing their preferred migration city. The suitability of their jobs with short distance location of work from their homes are among the strongest reasons that made them stay. Overall, the average living period of rural migrants is 3.5 years in Beijing and it is quite opposite from the previous research that rural migrants only stay in urban area less than 1 years (Zhou, 2002; Zhu, 2007).

5.3.1.5 Geographic Distribution

When considering whether or not they should migrate to Beijing, the distance from the migrants' province of origin is a notable factor. The five most common provinces are closest to Beijing. The survey shows that 80.0% of the rural migrants migrated from Hebei province, Shanxi province and Shandong province. 12.0% of the rural migrants are from Henan and Liaoning provinces. The distance from Hebei province, Shanxi province and Shandong province to Beijing only less than 500 kilometers; the distance from Henan and Liaoning provinces to Beijing also less than 700 kilometers (refer to Table 5.8).

Table 5.8 The distance of rural migrants' hometown to Beijing

Name of province	Frequency	Percentage(%)	Distance (km)
Hebei	320	80.0	Less than 500
Shanxi			
Shandong			
Henan	48	12.0	Less than 700
Liaoning			
Other provinces	32	8.0	More than 800
Total	400	100.0	—

Source: Wang. Social movement and re-structure, 1995.

In the survey, the rural migrants mentioned that they were drawn to move by the short distances from their hometown. As the previous researchers Li (1991) and Todaro (1969) have mentioned, rural migrants prefer to migrate to the nearest big cities. In this vein, this research has produced the same conclusion (see Fig. 5.7).

5.3.1.6 Household Size and Children Education

In their study of urban migration in Shanghai, Guangzhou and Hong Kong, Li (2000) has found that rural migrants prefer to bring their family to the cities. In this research, the outcome obviously follows their conclusion. As shown in Table 5.9, the most dominant household size is 4. And the rural migrants' average household size is around 4. However, the average household size in Beijing is around 3 (Wu, 2010). Therefore, it represented that the rural migrants have the bigger household size and it also could result in overcrowding (see Table 5.9).

The Housing Condition of Rural Migrant in Bejing Fengtai District

Fig. 5.7 Top five provinces for Beijing migration. Note that more saturated areas (in red and yellow) indicate a higher percentage of migrants to Beijing

Table 5.9 Household size of the rural migrants in Beijing   (%)

Household size	Frequency	Percentage
1	71	17.8
2	53	13.2
3	80	20.0
4	110	27.5
>4	86	21.5
Total	400	100.0
Average household size	Around 4.1	

Since most migrants do not possess any Beijing *hukou*, their children are restricted to go to public schools while private schools are unattendable due to higher school fees. As a result, 40.3% of the rural migrants decided to let their children to quit school. Instead of being educated properly, their children end up helping them or working with them in their small businesses (Wang, 2007). Due to without good education, also the government does not pay attention to these rural migrants. Therefore, their children's future development definitely will be affected(see Table 5.10).

Table 5.10 Rural migrants' children attend school in Beijing (%)

Children's education situation	Frequency	Percentage
Attend school	64	16.0
Not attend school	161	40.3
No children or not bring children to Beijing	175	43.7
Total	400	100.0

Therefore, having large families means heavy burden to the rural migrants. Without good education, the rural migrants could only do small business or find some low income jobs like hawker, petty trading etc. The researcher also noticed that some rural migrants sell fruit and vegetable beside the street, but sometimes with very few purchasers. It represented that the rural migrants' living conditions in Beijing is quite hardships.

5.3.1.7 Education Level and Types of Jobs

Through Pearson correlation analysis, the Sig. (2-tailed) 0.000 less than 0.05 shows that education level and types of job has statistically significance and could discuss by cross-tabulation(see Table 5.11).

Table 5.11 Rural migrants' education level and types of jobs (%)

Education level	Types of Job	Frequency	Percentage
No formal education	cleaner, hawker etc.	105	26.3
Primary school	hawker, petty trading etc.	100	25.0
Secondary school	hawker, petty trading etc.	100	25.0
Diploma	bus driver, technician etc.	40	10.0
Degree	work in company	30	7.5
Postgraduate	work in company or department	25	6.2
Total	—	400	100.0

(Pearson Correlation: Sig. (2-tailed) 0.000 < 0.05)

As the previous researchers Feng (1997) and Lu (2005) have mentioned, rural migrants' low education level result in the job is low level. In this research, the same result has been established. More than half (76.3%) of the rural migrants work in the low level job which is cleaner, hawker, petty trading etc. Wang (2007) and Portes (1995) whose studies are based on Guangzhou and New York mention that, rural migrants have no choice but to choose the low level job due to their low education levels. Therefore, the relationship between the two variables (education level and types of job) can have an impact on each other among the rural migrants.

5.3.1.8 Summary

Overall, the target of rural migrants in Beijing is to obtain higher income and increase their living condition. In this research, a majority (65.0%) of the rural migrants are male and below 42 years old. However, due to their low education level and unstable job, most of the rural migrants' income (84.5%) is lower than the average income in Beijing 5 223 RMB (BMBS, 2013). As they had a lack of good economic support, the lack of income and other economic supports see that 84.0% of the rural migrants' children are not attending proper schools, either public or private. The average time of stay in Beijing is 3.5 years for rural migrants and 10.8% of the rural migrants came to Beijing 10 years earlier. Obviously, the distance is an important factor for them to migrate. 92.0% of the rural migrants are from the rural areas near Beijing like Hebei, Shandong, Shanxi, Henan and Liaoning and they prefer to move and earn their living in the nearest big cities.

5.4 Housing Characteristics

Understanding about the housing characteristics could help us better understand the rural migrants' lives in Beijing. Based on the previous research, due to the low economy the rural migrants prefer to rent the small houses (usually less than 50 m^2) and with old facility. Also, many rural migrants share the house with other families. Thus, the overcrowding and privacy are affected. In Beijing, the overcrowding, housing privacy and housing facility problems stand out to be the crucial housing issues for rural migrants (Guo, 2006; Qian, 2003; Wang, 1995)(see Table 5.12).

Through Independent T-Test, Correlation analysis and Multi-nominal regression analysis, the above independent variables and dependent variable are tested to accept or reject the statistically significance between each other. If the Sig. more than 0.05, it will reject the statistically significant between the variables.

Table 5.12 Test the statistically significance of Independent and dependent variables

Independent variable		Dependent variable	Statistic analysis	Statistically significance
Overcrowding	Household size	Living condition	Independent T-Test	Accept
	Person-Per-Room		Independent T-Test	Accept / Reject
	Housing design		Correlation analysis	Accept
	Average living space		Correlation analysis	
Privacy	Housing function (living/work)	Living condition	Correlation analysis	Accept
	Roommate action		Correlation analysis	Accept
	Neighborhood action		Independent T-Test	Accept
	Public facility/ transportation		Independent T-Test	Accept
Facility	Housing facility	Living condition	Multi-Nominal regression	Accept
	Home appliance		Correlation analysis	Accept
	Social facility		Multi-Nominal regression	Accept

(Sig. > 0.05 will reject the statistically significant between the variables)

5.4.1 Overcrowding

For housing overcrowding measurement, the "Traditional and Alternative Definitions of Overcrowding" (Blake, Kellerson and Simic, 2007) use Persons-Per-Room (PPR) or Persons-Per-Bedroom (PPB) not more than one person is used to define housing overcrowding. However, due to the high living density and housing shortage in Beijing, Wu (2010) uses the measurements that are suitable to the real situation in Beijing which include Persons-Per-Room (PPR) or Persons-Per-Bedroom (PPB) all of which should not be habitated by more than two people or it will be defined as overcrowding(see Table 5.13).

Table 5.13 Persons-Per-Room (PPR) in Beijing Fengtai District

Persons-Per-Room (PPR)	Frequency	Percentage
1	11	2.75
2	76	19.0
3	115	28.75
4	137	34.25
> 4	61	15.25
Total	400	100.0

Note: Persons-Per-Room (PPR) or Persons-Per-Bedroom (PPB) more than two people is overcrowding, Wu, 2010.

As shown in Table 5.13, the most dominant group for Persons-Per-Room (PPR) is 4, which indicates that the room is highly congested. During the survey, the researcher noticed that most of the rural migrants' houses have only one or two rooms and with small sizes (less than 10 m^2). But most of the rooms are at least housed more than two persons, thus the overcrowding problem is very common among rural migrants. Surprisingly, 15.25% of the rural migrants share a room which occupies more than 4 people. And most of these people are friends or their relatives. Overall, 78.25% of the rural migrants share one room with more than two people. As Freedman (1975) and Rapoport (1976) mention, the rural migrants' housing overcrowding problem is very common in big cities especially among developing countries. Due to a small room (less than 10 m^2) but live more than two or three persons, many rural migrants complained that they fell uncomfortable and cannot sleep well due to the overcrowded room. As many people have to be 'sandwiched' inside one room, the housing overcrowding problem is very serious for rural migrants in Beijing.

Through Pearson correlation analysis, the Sig. (2-tailed) 0.006 less than 0.05 show that housing ownership and length of Stay in Beijing has statistically significance and could discuss by cross-tabulation. Based on the research, 90.5% of the rural migrants rent house in Beijing Fengtai District. However, only 9.5% of the rural migrants purchase house. The researcher found that majority of these rural migrants who purchase house in

Beijing are living here more than 5 years. While, 56.0% of the rural migrants purchase house that they already live in Beijing more than 10 years. In the survey, the researcher found that these migrants who purchase house in Beijing are belong to high education level (Postgraduate). And majority of these people are working in the company or government department(see Table 5.14).

Table 5.14   Housing ownership and length of stay in Beijing of rural migrants

Year	Buy house	Rent house	Frequency	Percentage
Less than 1 year	0	136	136	34.0
	0%	100%		
1 ~ 2 years	3	98	101	25.3
	3%	97%		
2 ~ 5 years	5	75	80	20.0
	6.25%	93.75%		
5 ~ 10 years	6	34	40	10.0
	15.0%	85.0%		
10 years before	24	19	43	10.8
	56.0%	44.0%		
Total	38	362	400	100.0
	9.5%	90.5%		
Average year	3.5 years			

(Pearson Correlation: Sig. (2-tailed) 0.006 <0.05)

In addition, according to the Beijing Municipal Bureau of Statistics (BMBS, 2011), the average housing size is 29.4 m^2. Based on this government report, the average housing size less than 29.4 m^2 will be define as overcrowding. Thus, the average housing size less than 29.4 m^2 defined as overcrowding will be used in this research(see Table 5.15).

Table 5.15 Rural migrants' average housing size in Beijing Fengtai District

Rural migrants' average living space (m^2)	Frequency	Percentage
< 10	168	42.0
11 ~ 20	95	23.75
21 ~ 29.4	60	15.0
29.4 ~ 40	57	14.25
> 40	20	5.0
Total	400	100.0

Note: The average housing size less than 29.4 m^2 will be defined as housing overcrowding, Beijing Municipal Bureau of Statistics, 2011.

Table 5.15 further shows that 80.75% of the rural migrants' average living space is less than 29.4 m^2, which represent the fact that their houses are overcrowding. This result produces the same conclusion with Liu (2007) and Luo (2009) where rural migrants' houses are always crowded. Only 19.25% of the rural migrants' living space is more than 29.4 m^2 and these people have migrated to Beijing for more than 5 years according to the survey. Overall, the majority of the rural migrants live in the crowded houses in Beijing. Similarly, the housing overcrowding is an important factor which affects other people's privacy (Altman, 1996; Freedman, 1971; Loo, 1973).

5.4.2 Privacy

In their study of rural migrants in urban areas, Qian (2003), Westin (1970) and Wang (2006) have found that rural migrants' housing privacy could be affected by their housing function, inside and outside their houses. In the survey, more than half (60.0%) of the rural migrants use their houses as places to stay and also to work. And many of them mentioned that they use their house to cook food and sell in the outside. Some of them said they made some handicrafts, furniture or something others in their house to be sold. Also, the researcher found that many of their friends or colleagues live together which might have resulted the privacy problem. More than 90.0% of them mentioned that their privacy were disturbed by work or business. They could not sleep well and they were always interrupted by customers or other roommates. Thus, their personal affairs and work are seriously affected inside the house (see Table 5.16).

Chapter 5 — Analysis and Results

Table 5.16 Function of rural migrants' houses

Housing function	Frequency	Percentage
Residential solely	160	40.0
Residential and work	240	60.0
Total	400	100.0

Although, 40.0% of the rural migrants used their houses solely for residential purposes, but still more than half of them complained that their privacy was affected by roommates or neighbors both inside and outside their houses. Due to the fact that houses are shared by other people (friends or relatives), the roommates always disturb their privacy like sleeping, writing something, making phone calls or others.

Through Independent T-Test, the Sig. (2-tailed) all less than 0.05 show that roommates, customer, neighbor, other people around house, public facility and transportation all have statistically significance to rural migrants' privacy (life and work) and could discuss by cross-tabulation.

Human's privacy was affected inside or around the house (Chan, 1997). Outside the houses, still more than two thirds of the rural migrants mentioned that their life and work were affected by neighbors or other people's actions like looking into the room through the window or always knocking on their doors to sell something. Due to the privacy problem, they have to draw the curtain even in the daytime. Also, sometimes the sellers knock their door two or three times and it serious affect their life. More than half of the rural migrants complained about the sounds of the taxis and buses around their houses, and their personal items like dry clothes, children playing outside or shopping goods were always spotted by the passengers or drivers. They also mentioned that many people living around them always threw rubbish or parked their cars in a disorderly manner, and these seriously affect their regular lives (see Table 5.17).

Table 5.17 Ranking of disturbance to rural migrants' privacy (life and work)

Ranking	Item	Mean	Independent T-Test
1	Roommates/customers	3.87	0.000
2	Neighbors/other people around house	3.65	0.006
3	Public facility/transportation	3.30	0.012

(Independent Sample Test: Sig. (2-tailed) < 0.05)

(Likert Scale, "Mean" more than 3 is positive)

Obviously, the disturbance to rural migrants' privacy (life and work) inside their houses was ranked at number one. Previous researchers Rodin (1976) and Sherrod (1974) state that the migrants' personal privacy inside their house is always affected, and this research follows the same result. For their neighbors or other people's disturbance, the negotiation should be used. However, for the public facility or transportation, as Beijing is a heavy traffic city, this situation cannot be avoided(see Table 5.18).

Table 5.18 The rural migrants' privacy was disturbed in their room

Privacy disturbed in rural migrants' room	Frequency	Percentage
Yes	357	89.25
No	43	10.75
Total	400	100.0

89.25% of the rural migrants mentioned that their personal privacy was affected inside the room by their roommates or other people. Also, they had mentioned that their privacy was disturbed as their houses are very crowded, meaning that there are too many people living together. This result is in tandem with the result derived by the previous researchers (Altman, 1996; Lu, 2010; Logan, 1993) where it is found that overcrowding is the precursor to privacy disturbance.

Overall, around 90.0% of the rural migrants' privacy was always disturbed either inside or outside their houses. And they also complained that their sleeping, chatting, working or even changing clothes always affected by the privacy problem. Thus, their life and work are seriously affected. To avoid the privacy disturbance entirely, the housing overcrowding should be mitigated in Beijing.

5.4.3 Facility

Access to housing facility is a crucial element for rural migrants in Beijing as it may be expected to determine their financial situation. Barber (2007) and Ding (2005) have concluded that rural migrants' poor housing facility in urban area due to their financial limitation. Due to the poor finance that restricted them to pursue the better housing facility. Liu (1995) carries out a research in India and he proves that finance is related to rural migrants' housing facility in urban areas. Therefore, to determine the rural migrants' housing facility status, how much percentage of the rural migrants have it is an important factor to measure it (Yang & Yan, 2000; Tolley, 1991).

Through Multi-nominal regression (Likelihood Ratio Test), the Sig. is 0.000 less than 0.05 shows that housing facility and household size has statistically significance and could discussed by cross-tabulations.

Beijing is the capital of China, with the basic housing facility like pipe gas (98.5%), running water (96.25%) and electricity ((95.0%) are satisfactorily provided to rural migrants. 82.5% of the rural migrant houses in the study were heated during the winter time in which the temporary could fall to below freezing. In the survey, the rural migrants mentioned they use the electrical heat in winter. However, only 66.25% of rural migrants' houses provide hot running water to take shower in their houses in winter. However, the remainder of 33.75% of rural migrants needs to go to the public bath house around their houses in winter. The public bath house is everywhere in China, the residents just pay fee and could take shower inside. Many rural migrants complain that they could not afford it due to the fee is significantly high. Therefore, the hot running water supply to migrant houses should be improved in Beijing.

According to the result, the researcher noticed that the basic housing facility like pipe gas, running water, electricity and housing heater, can be found mostly in the households with 4 or higher members. Nevertheless, these households are in need of such facilities since it was noticed that the households are almost always consist of children or old occupants. However, for other non-basic facilities such as balcony, private kitchen and entertainment space, singular household dominates the most. Normally, these are higher income households who have stayed in Beijing for longer periods and have already obtained a *hukuo*. In such cases, many of them mentioned that they are staying in Beijing on a permanent basis(see Table 5.19).

Table 5.19 Rural migrants' housing facility and household size in Beijing Fengtai District

Ranking	Item	Household size (frequency)					Frequency	Percentage
		1	2	3	4	> 4		
1	Pipe gas	69	50	79	110	86	394	98.5
2	Running water	66	52	80	105	82	385	96.25
3	Electricity	68	50	75	107	80	380	95.0
4	Housing heater	60	43	70	103	54	330	82.5
5	Hot running water	30	41	55	80	59	265	66.25
6	Balcony	65	40	52	55	20	232	58.0
7	Private kitchen	67	35	50	27	17	196	49.0
8	Entertainment space (living room)	50	23	26	15	16	130	32.5

(Multi-nominal regression (Likelihood Ratio Test), the sig. is 0.000 < 0.05)

Only 58.0% of rural migrants' houses have a balcony, so the remaining 42.0% of rural migrants need to find an appropriate place to dry clothes inside or outside their houses. For the kitchen, the majority of rural migrants need to share the kitchen with other people. However, they only need a place to cook, so they do not care if the kitchen is private or not. As Guo (2004) and Zhang (1997) conclude, rural migrants' living requirement is very low in urban areas. The majority of the rural migrants plan to go back to their hometown, therefore, the living requirements for the temporary place in which they stay in Beijing are reduced or downgraded. Very few (32.5%) rural migrants' houses have an entertainment place. This result automatically confirms the rural migrants only pay attention to the basic housing facility (see Table 5.20).

Table 5.20 **Rural migrants' home appliances in Beijing Fengtai District**

Ranking	Item	Frequency	Percentage
1	Refrigerator	372	93.0
2	Washing machine	367	91.75
3	Television	285	71.25
4	PC/laptop	220	55.0
5	Air-Conditioner	216	54.0
6	Others (e.g. microwave)	167	41.75

As the previous researcher (Bian, 2005) have mentioned, rural migrants' living requirement is very low. Nevertheless, this study has shown somewhat different findings. Refrigerator and Washing Machine are the bare necessities of life today. Thus, these home appliances are mostly owned by rural migrants in Beijing. In the survey, the rural migrants mentioned that they needed the refrigerator to keep their food. They also needed the washing machine to wash clothes every day. However, television, PC, laptop, Air-Conditioner etc are considered as non-necessities for the migrants, but still more than half of the rural migrants have them. It came out some difference that the home appliances are very low (less than 30.0%) for rural migrants in urban area (Bian, 2005). In the survey, many rural migrants mentioned that they purchase the second-hand home appliances and the price is very cheap. Also, due to the houses are rented from the local residents and already fully-furnished. So, the ownerships of these home appliances are very high for rural migrants in Beijing.

Chapter 5 Analysis and Results

Through Multi-nominal regression (Likelihood Ratio Test), the Sig. is 0.000 less than 0.05 shows that social facility and household size has statistically significance and could discussed by cross-tabulations. According to Wu (1996), the migrants are only concerned about the basic necessities in life which is shown in their locational choice of dwelling units. 96.25% of the rural migrants in Beijing chose their houses as nearby to supermarket as possible to enable them to purchase their basic daily household supplies such as fruits and vegetables. For the garden and resting place, 72.5% and 71.25% of the rural migrants hope to have them around their houses. Thus, they could take a walk or joggling in the places. For some high cost entertainment venues like the cinema, hotel, KTV etc., more than half of the rural migrants did not care much about their existence in the areas. But in the survey, many rural migrants said they do not wish to live around the cinema, KTV etc due to it affect their children's education. Also, they feel that it will negatively influenced their life(see Table 5.21).

Table 5.21 Rural migrants' social facility and household size around their houses in Beijing Fengtai District

Ranking	Item	Household size (frequency)					Frequency	Percentage
		1	2	3	4	> 4		
1	Supermarket	62	45	69	106	103	385	96.25
2	Garden	55	30	60	101	44	290	72.5
3	Rest place	57	38	57	95	38	285	71.25
4	Restaurant	69	46	50	37	37	239	59.75
5	Cinema	65	50	47	33	38	233	58.25
6	Hotel	67	39	49	35	35	225	56.25
7	Others (e.g. KTV)	63	33	41	20	16	173	43.25

(Multi-nominal regression (Likelihood Ratio Test), the sig. is 0.000 < 0.05)

Note: 2 kilometers around their houses.

Source: Beijing Construction Department. Beijing Construction Department report: Rural migrant housing facility in Beijing conducted in 2006.

From the result, the researcher found out that most of the family with 4 household members lives closest to the supermarket, garden, rest place around their houses. The rural migrants mentioned that especially the family with children or old people, they need the supermarket to shopping and the garden or rest place to take a walk. Also, singular household seemed to locate nearest to the "luxury" facilities like the restaurant, cinema, hotel and KTV. Due to these people are single in Beijing and majority of them belong to high income, obviously they need these entertainment places to enjoy their life in Beijing.

5.4.4 Summary

Due to the large population and the limited land, Beijing quickly becomes a highly densely-populated city. Therefore, the housing overcrowding problem is most common among rural migrants. Also, their housing privacy is affected by housing overcrowding to a large extent. For rural migrants' housing facility, more than 90.0% of them only considered their basic housing facility. Therefore, to improve their housing facility, not only that their finance can be improved, but also their lives can be enhanced as well.

5.5 Influence on Living Condition

5.5.1 Living Overcrowding

Wu (1996) carries out a research in Germany and he concludes that overcrowding could seriously influence people's living condition. Wang (2003) who has performed the research in Beijing and Shanghai has also proven this result. This research testifies the result again that housing overcrowding could influence rural migrants' living conditions like sleeping, working and many others.

In this research, 27.0% and 28.0% of the rural migrants "Strongly Agreed" and "Agreed" that their living condition was seriously affected by housing overcrowding. This result (55.0%) shows the majority of rural migrants did not want their living condition to be affected by overcrowded houses. However, 15.0% and 11.25% of the rural migrants marked the "Disagree" and "Strongly Disagree" columns to this result. During the survey, these people mentioned that due to their poor finance, they had no choice but to live in such an overcrowding area. Overall, at least more than half (55.0%) of rural migrants' living condition is suffered by the housing overcrowding issue. Thus, to increase rural migrants' living condition in Beijing, housing overcrowding should be avoided (Vogel, 1990)(see Table 5.22).

Table 5.22 Many people live in one house with you could disturb living condition like sleeping, working and others

Item	Frequency	Percentage	Score
Strongly agree	108	27.0	540
Agree	112	28.0	448
Neutral	75	18.75	225
Disagree	60	15.0	120
Strongly disagree	45	11.25	45
Total	400	100.0	1 378
Mean		3.45	

Through Independent T-Test, the Sig. (2-tailed) all less than 0.05 show that sleep, study, working condition, recreational activities and household relations all have statistically significance to housing overcrowding and could discuss by cross-tabulation. Ding (2005) state that migrants tend to pay attention to their basic living condition in urban areas. This research in Beijing produces the same result. Sleep is ranked as the first condition that was affected by housing overcrowding and this is of serious concern among the residents. As some rural migrants mentioned in the survey, they did not want their sleep to be disturbed by noises or discomfort caused by overcrowding. Study and working conditions were also affected by this. However, recreational activities and household relations were not very important to rural migrants in Beijing. Even household relationship produce a negative result, which means housing overcrowding cannot affect their household relationship. They also mentioned in the survey that household relations were not important to them. More than half of the rural migrants live in a house for a long time, but apparently they even have not talked to each other although they are in the same houses. Therefore, the rural migrants live in the houses are treated as covering their basic needs such as sleeping and resting since they are outside the house most of their time (Wang, 2007), the relations (communicating, discuss etc.) between members are not considered as important because of this lifestyle (see Table 5.23).

Table 5.23 Ranking of housing overcrowding to affect rural migrants' working, recreational etc.

Ranking	Item	Mean	Independent T-Test
1	Sleep	3.96	0.000
2	Study	3.67	0.000
3	Working condition	3.65	0.000
4	Recreational activities	3.10	0.010
5 (Negative)	Household relations	2.87	0.037

(Independent Sample Test: Sig. (2-tailed) < 0.05)
(Likert Scale, "Mean" more than 3 is positive)

Because the rural migrants do not consider about the relationship between other people in their houses due to the priorities in life, as a consequence, some problems are bound to happen. Many rural migrants mentioned that most of the disturbances happened in the early morning and at night. In the early morning, due to the fact that too many people hurry to go to work, the bathroom will have to accommodate many people at the same time. So, sometimes the people will tend to quarrel over who uses the bathroom first. At night, some people would come back late with loud noise, and this makes other people angry. Some other people want to do some work or sleep in their bedroom, and the quarrel or even fighting will disturb them. Sharing kitchen and living room also invites bickering and quarrelling (see Table 5.24).

Table 5.24 Rural migrants' living condition (e.g. sleeping, cooking, working etc.) was Influence by their housing overcrowding

Item	Frequency	Percentage	Score
Strongly agree	135	33.75	675
Agree	128	32.0	512
Neutral	56	14.0	168
Disagree	53	13.25	106
Strongly disagree	28	7.0	28
Total	400	100.0	1 489
Mean		3.72	

Overall, the rural migrants' living condition (e.g. sleeping, cooking, working etc.) was seriously influenced by housing overcrowding. 33.75% and 32.0% of rural migrants agreed that their living condition was indeed, affected by their housing overcrowding. The result confirms the previous research that housing overcrowding affects migrants' living condition. To improve rural migrants' living condition in Beijing, not only housing overcrowding needs to be avoided, but their living condition standard should also be improved.

5.5.2 Living Privacy

Chan (1997) in his study mentions that migrants had less adequate privacy due to the overcrowded houses in urban areas. Thus, their living condition was affected by housing privacy and related issues. Epstein (1994) has also stated in his research that rural migrants' living density resulted in problems related to housing privacy. Without good housing privacy, their living condition will definitely become a critical problem.

Through Independent T-Test, the Sig. (2-tailed) all less than 0.05 show that roommates, customer, neighbor, other people around house, public facility and transportation all have statistically significance to living condition (life and work) and could discuss by cross-tabulation.

As the migrants' housing privacy always affects their living condition, this research mainly focuses on rural migrants' roommates, customers (using own house to do business), neighbors or other people and the public facility or transportation. In sum, the rural migrants had been positive that the lack of housing privacy could seriously affect their living condition (see Table 5.25).

Table 5.25 Ranking of rural migrants' housing privacy to influence living condition (life and work)

Ranking	Item	Mean	Independent T-Test
1	Roommates	4.10	0.000
2	Customers	3.78	0.000
3	Neighbors/other people around house	3.60	0.003
4	Public facility/transportation	3.52	0.010

(Independent Sample Test: Sig. (2-tailed) < 0.05)
(Likert Scale, "Mean" more than 3 is positive)

The roommates (family members, friends or colleagues) take up the first rank as contributing to the poor privacy that further affects the migrants' living condition. As the rural migrants had mentioned, their inadequate housing privacy seriously affects their living condition, thus their life and work cannot operate normally. They also stated that their housing privacy tended to be affected by their housing overcrowding. This result confirms that overcrowded houses affects housing privacy, thus making living condition deteriorate further (Li, 2004; Li, 2006).

Also, this research has shown that rural migrants' living condition affected by internal housing privacy is very high. But outside their houses, the living condition influenced by housing privacy decreases. This result confirms the findings by previous researchers Sababu (1998). Also, the result shows that rural migrants pay more attention to their life inside the four walls of their houses. Therefore, to improve rural migrants' living condition in Beijing, their housing overcrowding and housing privacy problems have to be addressed (see Table 5.26).

Table 5.26 Rural migrants' living condition (e.g. sleeping, cooking, working etc.) was influence by their housing privacy

Item	Frequency	Percentage	Score
Strongly agree	127	31.75	635
Agree	115	28.75	460
Neutral	75	18.75	225
Disagree	60	15.0	120
Strongly disagree	23	5.75	23
Total	400	100.0	1 463
Mean		3.66	

Overall, the rural migrants' living condition (e.g. sleeping, cooking, working etc.) was seriously influenced by housing privacy. 31.75% and 28.75% of rural migrants were positive that their living condition was affected by their housing privacy. The result confirms the previous research that housing privacy seriously affects migrants' living condition. Therefore, to improve rural migrants' living condition in Beijing, their housing privacy problems should be minimized.

5.5.3 Living Facility

In their research Wang (1999) and Zhan (2003), maintain that housing facility is much related to residents' living condition, and this is particularly so for migrants. Good living condition should indicate a good better housing facility. However, in this current research about rural migrants in Beijing, the result suggests a little difference.

Through Independent T-Test, the Sig. (2-tailed) all less than 0.05 show that kitchen facility, bathroom facility, home appliance, balcony, recreation facility near houses and entertainment place inside houses all have statistically significance to rural migrants' living condition (life and work) and could discuss by cross-tabulation (see Table 5.27).

Table 5.27 **Ranking of rural migrants' housing facility to influence living condition (life and work)**

	Ranking	Item	Mean	Independent T-Test
Positive	1	Kitchen facility	3.75	0.000
	2	Bathroom facility	3.60	0.000
	3	Home appliances	3.53	0.000
	4	Balcony	3.10	0.007
Negative	1	Recreation facility near houses	2.63	0.010
	2	Entertainment place inside houses	2.80	0.015

(Independent Sample Test: Sig. (2-tailed) < 0.05)
(Likert Scale, "Mean" more than 3 is positive)

The result represents that rural migrants pay more attention to their kitchen facility, bathroom facility and home appliances. As they mentioned, their lives cannot be devoided of these housing facilities. The kitchen facility is ranked number one, where more than half of the rural migrants said they need the kitchen where they cook every day. Also, the bathroom facility is the second most important housing facility as it is also a basic necessity. Meanwhile, not many rural migrants mind if their houses have balcony or not as it is not very important to their lives. As some rural migrants had stated, they could dry their clothes inside their houses. Thus, is not necessary to have a balcony. This result follows the previous researcher whereby rural migrants only emphasises the basic housing facility to improve their living condition (Wang, 2007; Zhou, 1998).

However, this research in Beijing came out a little different in the sense that poor recreational facilities near their houses and not having entertainment place inside their houses all cannot affect their living condition. During the survey, majority of the rural migrants' said they did not need the recreation facility near their houses. Also, for the entertainment place inside their houses like a living room, they stated that it was still not necessary. This result conflicts with the majority of the previous researchers stating that entertainment facilities still influence rural migrants' living condition in urban cities and represent the fact that the rural migrants' living condition requirement in Beijing is still very low (see Table 5.28).

Table 5.28 Rural migrants' living condition (e.g. sleeping, cooking, working etc.) was influence by their housing facility

Item	Frequency	Percentage	Score
Strongly agree	135	33.75	675
Agree	121	30.25	484
Neutral	79	19.75	237
Disagree	42	10.5	84
Strongly disagree	23	5.75	23
Total	400	100.0	1 503
Mean		3.76	

Although the majority of rural migrants have adopted a negative attitude on the impact of the entertainment facility on their living condition, the overall result is still positive. 33.75% and 30.25% of the rural migrants had ticked "Strongly Agree" and "Agree" on the fact that their living condition was influenced by their housing facility. This result strongly confirms that to have good living condition, housing facilities must be improved for rural migrants.

Overall, due to rural migrants' economic problem and low living requirement, they only pay attention to the influence of the basic housing facility on their living condition. However, in order to improve the living condition for this group of people in the cities, their housing facility improvement is regarded as necessary.

5.5.4 Summary

Through Independent T-Test, the Sig. (2-tailed) all less than 0.05 show that housing facility, housing overcrowding and housing privacy all have statistically significance to rural migrants' living condition (life, work, entertainment etc.).

Comparing the housing overcrowding, housing privacy and housing facility in terms of their influence on the rural migrants' living condition in Beijing, the housing facility is ranked number one. Due to these three housing characteristics they are serious issues of rural migrants' housing in Beijing. Therefore, to improve the living condition of this group, the three housing characteristics must be revisited and improved.

5.6 Conclusions

It is apparent that the majority of rural migrants in Beijing Fengtai District covered by the study have chosen housing facility as the most dominant problem which influences their living condition. Majority of the rural migrants in Beijing considered housing facility as the most important issue in their lives and this result follows the last result (see Table 5.29).

Table 5.29 **Ranking of rural migrants' housing overcrowding, privacy and facility to influence their living condition (life, work, entertainment etc.)**

Ranking	Item	Mean	Independent T-Test
1	Housing Facility	3.81	0.000
2	Housing Overcrowding	3.72	0.000
3	Housing Privacy	3.66	0.005

(Independent Sample Test: Sig. (2-tailed) < 0.05)

(Likert Scale, "Mean" more than 3 is positive)

Meanwhile, 30.25% of rural migrants considered overcrowding as more important than housing privacy (27.0%) in influencing their living condition (Fig. 5.8). As the previous researchers have mentioned housing overcrowding is directly linked with the lack of housing privacy (Baldassare, 1999; Mcgrew, 1970). This result clearly follows their result. Therefore, to solve the rural migrants' housing privacy problem, their housing overcrowding problem first needs to be settled.

Although housing facility, overcrowding and privacy are three grave housing issues

The Housing Condition of Rural Migrant in Bejing Fengtai District

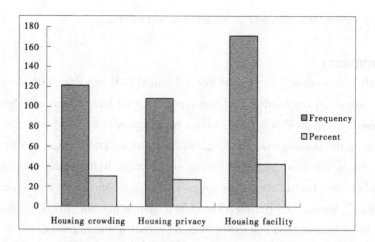

Fig. 5.8 Compare the housing characteristics to influence living condition

that can affect rural migrants' living condition, more than half of rural migrants decided not to improve or not to move someplace else altogether. 28.75% and 27.25% of rural migrants in Beijing Fengtai District still live houses with the existing housing characteristics. As these rural migrants had mentioned, Beijing is their temporary home. Finally, they will bring money and return to their hometown. So they had decided not to change their current housing characteristics. This result are in line with the previous researcher who states that the city is the rural migrants' temporary living place, ultimately, the majority of them will go back to their hometowns (Wu, 2010) (see Fig. 5.9).

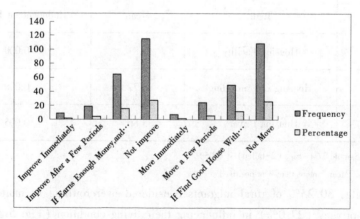

Fig. 5.9 Rural migrants' future action in regards of housing characteristics

However, still have more than one quarter of rural migrants plan if earns enough money then improve it (16.25%) or if find good house with cheat rental then move (12.5%). It represents that some rural migrants still plan to pursue a good life in Beijing (Cai, 2003). Very few rural migrants plan to improve their lives immediately (2.25%) or move immediately (2.0%). In the survey, two of these rural migrants stated that their salaries were high and it was enough for them to afford better houses. They want to create good living condition for their children.

Nonetheless, the high-income rural migrants only occupy a very small proportion in this research. Therefore, to consider the housing characteristics and influence on their living condition, the majority of low-income rural migrants should be well accounted for in Beijing.

Chapter 6
Implication of Findings

6.1 Introduction

In assessing the implications of the findings on the rural migrants in Beijing, the chapter is divided into several sections. The first part deals with its implications in relation to the housing theories put forth. Further discussion is geared to the implications of government policy on migrant housing and *hukou* management. The final section examines the implications on the rural migrants.

6.2 Implications on the Housing Theories

6.2.1 Living and Housing Theory

There is an agreement amongst some of the proponents of living and housing theories (Havel, 1957) which implies housing overcrowding standard determination (Blake, Kellerson and Simic, 2007; McArdle & Mikelson, 1994). Researchers and scholars have indicated that the number of residents should not be more than the number of room or it will fall into the category of 'congested housing'. Wu (2002) stresses that two or three rural migrants living in one room will result in overcrowding and therefore, housing privacy will definitely be affected. However, the findings from this study present a somewhat different scenario.

It appears that although these researchers have stated that the number of residents should not be more than the number of room, but the outcome was obtained from developed countries thus, it was deemed as not quite suitable in developing countries, especially the ones with large population like China. While based on the real situation in China, especially in big cities like Beijing and Shanghai, the average population in one square kilometer is more than 30 thousand. So it is difficult to supply a single room for each resident, especially for the rural migrants with lower finance. In this research, two or three rural migrants squeeze into one room is very common in Beijing. Ironically, the rural migrants can accept this living condition. In the survey, many rural migrants mentioned that they leave their hometown and live in this remote city, with no relatives and friends. Even their family squeeze inside the overcrowded room, they can chat and talk as they wish. So, they do not feel lonely in Beijing. Even among Beijing's local residents, two children living in one room or one child and parents sharing one room is very common, and they do not feel that the room is too crowded. As a result, to define the housing overcrowding standard we should base this on real situations in different countries.

Also, the housing and living theory (Havel, 1957) defines the good housing should have a balcony, a garage, even a swimming pool. The definition is based on the developed countries like Germany, France, the UK or the US. It is impossible to define the good housing and used it in China. China was a developing country with large population. Only very few people in China could afford housing with a garage and swimming pool, while the majority of urban residents live in the apartments. If the housing has a kitchen, a living room, a bathroom and one or a few bedrooms, it is thus regarded as good housing in China. In this research, some rural migrants have to share the kitchen and bathroom, and even some rural migrants rent their houses without having a bathroom, but they do not complain too much. In the survey, the rural migrants mentioned that they do not feel lonely in the overcrowded room. They also mentioned that poor housing facility could encourage them to work and strive to improve their life. Therefore, due to the crowding situation with large population, the researcher use the definition that each room should not live more than two persons than the "Traditional and Alternative Definitions of Overcrowding" that each room should not more than one person (Blake, Kellerson and Simic, 2007). Thus, the housing and living theory (Havel, 1957) defines "adequate housing" at a very high level, and thus, some improvements should be improved in this theory(see Table 6.1).

Table 6.1 Compare the rural migrants' living condition between who get *hukou* and without *hukou*

Item	Get *hukou*	Without *hukou*
Very Good	16	7
Good	15	17
Neutral	7	315
Bad	2	11
Very Bad	—	10
Total	40	360

Overall, the housing and living theory (Havel, 1957) defines the living condition in the higher level. But due to majority of the rural migrants only plan to stay in Beijing for short time and they define their quality of life at a lower level. Based on Table 6.1, the majority of the rural migrants (without *hukou*), cannot purchase a house and without dependable income, but they still think their living condition in Beijing is "Neutral". Therefore, the majority of the rural migrants (without *hukou*) still consider their quality of life in Beijing is acceptable. Only very few rural migrants who "made it" in Beijing, who purchase the house, car, get the *hukou* (Beijing ID) and become the more successful group of people. The majority of them think their living conditions in Beijing is good. Therefore, the living and housing theory is more suitable among this small group people and who are successful in Beijing.

6.2.2 The Housing Development Theory

The housing development theory builds the basis of discussions on the premise of the relationship formed between the migrants and housing policy in China, also, it refer to the international like US, UK, Germany and France etc. It has been argued that once so many migrants have migrated into the urban areas, the next challenge is to solve the multifarious housing issues that will emerge. In the long run, modernization would eventually see the influx of more workers and small business persons to maintain the social activities. With a lot of migrants residing in the city, housing problems are becoming more and more crucial. Agreed by many researchers (Wu, 2010; Zhou, 2008), the housing development should pay attention to these groups of people. According to the housing theory (Hao, 2009) migrants would not disappear as a result of modernization and industrialization. In China, the urban areas would not be able to support so many migrants if the migration process continues. Therefore, the Chinese government should pay attention to these migrants.

This theory also emphasizes that due to the housing law in China, unlike the migrants, the urban local residents could enjoy the housing subsidy. This is a drawback in China housing system and thus, it can be improved like in some western countries, such as the US, Sweden and Holland in which the housing subsidy covers each all residents and migrants (Tian, 1998). The rapid increase of migrant population in cities like Beijing, has forced the city government to limit the granting of *hukou* to migrants who are able to occupy higher and more permanent level of jobs, but due to the limitation of their education levels, very few of them could get the higher level jobs. Thus, it is quite difficult for them to get the Beijing *hukou*. Therefore, the government should consider giving *hukou* to the migrants who get jobs at all levels matching it with the particular length of stay. Thus, the migrants could enjoy the housing subsidy in Beijing and their housing problems could be minimized.

Another unique characteristic of this theory is related to the statement of Hao (2009) whereby public housing policy is carried out in western countries like the US, the UK, Sweden, Holland and Singapore. The public housing is built for poor people and migrants. Currently in China, the government has built some low-rent housing only provided for urban local residents, except for migrants. The low-rent housing only needs to pay very low rental per month and can live in proper houses. Therefore, if the low-rent housing could cover the migrants in Beijing, the migrants' housing problem could be solved easily. Thus, this housing system should benefit migrants either and allow them to enjoy the low-rent housing as well. This can give more implications towards the housing policy in future and produce a harmonious society.

6.3 The Implications on Government Policy

6.3.1 Migrant Housing

The Chinese government, as the leadership of the Communist Party, has the power within the People's Republic of China and has the responsibility to maintain the Construction Department have included a series of measures to provide assistance to the urban local housing that covers the housing facilities, housing subsidies, housing management services as well as the housing marketing. Although the Chinese housing policy improves every year, however, compared to the urban local residents, the migrants still cannot enjoy these housing benefits. Due to too many migrants migrate to Beijing every year and the Beijing government cannot support housing to all the migrants. Also, the Chinese government wishes the migrants back to their hometown finally due to the heavy burden in the big cities. Thus the government strictly controls the *hukou* to the migrants. Thus, only the migrants who find high level jobs (government department or big company) and get Beijing *hukou*. Thus, migrants surge their way into the urban area and live in the poor housing without much support from the government. The only supportive is the government supplies some low level job for a few migrants in urban area.

Basically, there are three kinds of housing provided for urban local residents in China. They are commercial housing, economically affordable housing and low-rent housing. For local urban residents, they could choose housing based on their financial capabilities. However, migrants cannot enjoy any kind of housing facilities in Beijing. The only type of accommodation for migrants in Beijing is to rent houses or lives in some other places. Due to so many migrants migrate to the urban area and compete the social resources with the local residents. The relationship between local and migrants was deteriorated. The local government use *hukou* to separate the local and migrants, thus to give more benefits to local residents. But the migrants cannot enjoy any of these facilities. So, it is indicated here that the local government policy is not at par with the needs of the migrants. The housing policy, then, needs to treat each person as equal and not being biased towards them. If migrants could enjoy the housing benefits (economically affordable housing or low-rent housing) the same way as the urban local residents, the migrants' housing characteristics and living condition could be improved.

Another aspect of housing policy implication is the marketing of the commercial housing. If the housing policy could reduce the commercial housing price, many migrants with local ID could afford the commercial housing and live in the urban areas more easily. Thus, these migrants are willing to invest and this is good for urban economy development.

Finally, the Beijing government and Chinese Construction Department need to create a housing policy especially designed for migrants in future. Thus, these migrants could be helped easily when they are backed by a clearer and more just housing policy. Also, similar to the "public housing" policy in western counties, some neighborhoods could be built and provided for migrants only. Migrants live in these neighborhoods, not only could improve their housing characteristics and living condition, but also the government do the management easily. Unfortunately, in the semi-structure interview, the government officers mentioned that the government plan to create a policy especially for migrants only. But due to the migrants' population number increase too fast and the Beijing government has no ability to solve so many migrants' housing problem. Thus, this housing policy was not carried out by the government and they are stucked with the similar difficulties. This is a drawback and the government should care about the migrants as well in future and the housing policy should be justified towards them.

6.3.2 *Hukou* Management

Due to the large population in China, the Chinese government uses *hokou* to divide people into local and migrants. The local residents who hold the local *hukou* could enjoy the social welfare, but the migrants who do not have *hukou* cannot enjoy any social welfare. However, as it happens, many rural migrants migrate to big cities, to work and live there for several years. However, their *hukou* still belongs to their hometown, thus they cannot enjoy many benefits in the urban area. Due to too many migrants migrate to Beijing every year, the Beijing government cannot support the *hukou* to all the migrants. Due to the Chinese government wishes the migrants to be back to their hometown finally to reduce the heavy burden of the big cities. Thus, the Beijing government requires the migrants with higher level jobs in government department or big companies to acquire a *hukou*. Again, without a *hukuo* the majority of the migrants could not enjoy much social welfare in Beijing including housing, subsidy, education etc. The government only provides some training for the migrants and supply some low level jobs like cleaner, ticket seller etc. Therefore, the *hukou* management for the migrants (urban and rural migrants) should be revised properly.

Also, the Beijing government should improve the policy in the future to pay attention to the migrants' children, increase their chances to be educated etc. Thus, some special policy should be created for migrants. Like some public school could accept the migrants' children to attend school if they registered in the Beijing Migration Control Department. Also, their children could have the equal right to attend the college entrance examination in Beijing. Therefore, with the supportive of the government educational policy, the migrants' quality of life could be increased as well.

In China, many rural migrants migrate to the big cities like Beijing, Shanghai, Shenzhen and Guangzhou etc every year, but still very few of them could be granted the local *hukou*. Even after ten years living and working in Beijing, not more than 10% of the rural migrants could get Beijing's *hukou*. According to the investigation, many rural migrants have stated that the Beijing Migration Control Department does not give them the *hukou* due to the fact that they cannot find permanent jobs. Also, it is due to the low efficiency of the office together with the discrimination to the rural migrants. Therefore, the majority of the rural migrants who live and work in Beijing for more than five years still cannot get Beijing *hukou*. Thus, they cannot enjoy many social benefits in Beijing as compared to the local residents.

The research findings show that getting the Beijing *hukou* could help rural migrants to increase their living condition because they can also enjoy the social welfare in Beijing. By having the Beijing *hukou*, the rural migrants could enjoy the commercial housing, economically affordable housing and low-rent housing just like the locals. Thus, their housing conditions can be improved and overcrowding avoided, other than they will also gain access to privacy and good facilities.

Therefore, the Beijing Migration Control Department should revise the *hukou* policy. First, it may be possible to grant the migrants the Beijing *hukou* after they work and live in Beijing for more than one or two years, and to treat them as they do to the local residents. Secondly, the Beijing Migration Control Department should increase their working efficiency and without discrimination. Therefore, the migrants' housing characteristics and living condition could be improved if this issue of *hukou* is settled. Also, after they get the *hukou*, they are willing to improve their housing and living standard in future.

6.4 The Implications on the Rural Migrants

6.4.1 Rural Migrants' Characteristics

The characteristics of the Chinese rural migrants studied here appear to typify these people as having low income and low level education, migrating to big cities throughout the world. Their low level of education, however, shows that this makes it hard for them to find a good job with higher income. Also, due to without Beijing *hukou*, they cannot enjoy many social welfare in Beijing including the housing subsidy. In the survey, the migrants complained that they cannot enjoy any housing welfare in Beijing. At the same time, the low income forces them to rent the overcrowded housing with poor facilities. Therefore, their housing characteristics and living condition in Beijing are unsatisfactory and may lead to poor quality of life.

Nevertheless, among the rural migrants, many of them tend to bring their whole family to Beijing and do small businesses there. Since the majority of them were middle aged persons with many dependents (wife, children, parents etc.), the rural migrants carry great responsibilities to support their families. Coupled with their low level of education, they are forced to do some jobs with dirty and dangerous characteristics. In the survey, the researcher noticed that many rural migrants find the jobs as a street or bathroom cleaner, dredge sewer worker etc, which quite unsanitary. Many rural migrants mentioned they always sick due to the unhygienic work environment. Also, many migrants work in the construction sites and work high above the ground, which is quite insecure.

Previous studies on rural migrants have demonstrated that their wish to migrate to big cities is to earn money and to improve their living condition. However, due to their low education and difficulty in getting the urban *hukou*, their wishes were not realized. So it is very difficult for the rural migrants to find good jobs in big cities. Therefore, the rural migrants should be given help by the government. At the same time, the *hukou* and housing policy should be revised and without bias. Finally, government should also treat the locals and migrants (urban and rural migrants) fairly.

6.4.2 Savings

The savings is an important factor to stimulate the rural migrants' migration to big cities. The rate of savings depends on the rural migrants' education level, occupation and type of business investment. However, the majority of rural migrants have a low education level which translated into low income as a result of low level occupations. Even when they engage themselves in small businesses in big cities, the lack of enough investment and high cost of living always restrain their income.

The savings of rural migrants could improve their housing characteristics and living condition in big cities. However, most of the rural migrants plan to bring the money that they earn back to their home towns. In the survey, the rural migrants mentioned that they will bring the earn back to their hometown to build the new house there, also to improve the housing facility in that place. Moreover, some of them said they plan to buy a car or save the money for the children. So, they spend little money to improve their housing characteristics and living condition in these big cities. The majority of rural migrants consider the houses in big cities as only a temporary shelter, while the houses in their hometowns are their permanent places of abode. Therefore, the rural migrants' way of thinking further deteriorates their housing and living standards.

Getting the *hukou* or otherwise is also an important factor which influences the rural migrants to live in big cities on a temporary or permanent basis. Due to the large population in China and millions of migrants migrate to the urban area. The migrants compete the social resource with the local residents fiercely. Therefore, from 1949 (China establish year), the local government uses the *hukou* (local ID) to divide local and migrants strictly. The *hukou* protect the local residents' rights and migrants cannot enjoy it. This is a bias for the local residents. Only the migrants who find the high level jobs in government department or big company, then they could obtain the local *hukou* in the city. The social resource could only enjoyed by who hold the local *hukou*. But without local *hukou*, the migrants cannot enjoy a lot of social benefits in that city.

If the local governments give them the *hukou*, so they could enjoy the social welfare in big cities. But majority of the rural migrants cannot obtain Beijing *hukou* due to cannot find the high level jobs which in government department or big company. Thus, many rural migrants will decide to stay in big cities temporarily. They are not willing to spend capital on their houses and to increase their housing and living standards. If the *hukou* policies change and the government could give them *hukou* (local ID) if they could find a job whether in low, middle or high level, the rural migrants' housing characteristics and living conditions will definitely be increased.

6.4.3 Occupational Development

As has been discussed above, due to the rural migrants' low education level and poor skills to the migrants are restricted to find good jobs in big cities. Therefore, the local government should expose the rural migrants to some training on as for them to develop their social skills. Also, after training, the local government should give them some suitable occupations, for example bus driver, machine repairer, building worker etc. If big cities like

Beijing, Shanghai, Shenzhen and Guangzhou could provide good working opportunities to the rural migrants, the participation from the rural migrants will be very active.

Unfortunately, due to the large population movement in China, the local government cannot provide enough occupations to each migrant. Therefore, some special policy should be created to provide more opportunities to people, especially the disadvantaged people like the migrants. For example, one kind of housing should be created for migrants only and provide them a good shelter. Also, some public school should accept the migrants' children and allow them to attend the college entrance examination. Due to the severe occupational competition in big cities, the local government should pay more attention to care about the migrants in future.

Overall, the big cities could not have achieved high level of development without the migrants' help. Without the migrants (urban and rural migrants), for instance, the "2008 Olympics" organized by Beijing, "2010 Shanghai World Exposition" as held by Shanghai, and the "Asian Games" as organized by Guangzhou in 2010 etc. cannot be made possible. Many venues, stadiums and housing have been built by migrants (urban and rural migrants). They have undeniably given great contribution in the big cities' development. But due to their low education level and do the low level job, they are neglected in the city. Therefore, the government should pay more attention to these migrants and they should not be neglected.

This chapter has identified the broader implications of the research findings for the theory, policy and rural migrants. Apparently, the large population movement and rural migrants' low education level and poor social skills have further restricted their income. Therefore, their housing characteristics and living condition will be affected. The situation has been worsened by the discrimination of the local government and the difficulty to get the local *hukou*. Nevertheless, there has been some progress and it is possible to identify some pointers for the future. These are discussed in the final chapter.

Chapter 7
Overall Conclusion

7.1 Introduction

After analyzing the implication of the findings on the rural migrants' housing in Beijing Fengtai District, this chapter concludes the study by examining the relationship between housing characteristics and living condition. It also discusses the future of the housing system in Beijing, implying some repercussions that it has on the whole country. Finally, it touches on some other possible areas for future research.

7.2 Summary of Discussion

The main aim of this study was to examine a particular form of rural migrants by analyzing the linkages that were developed between the housing characteristics and their living condition, also investigating the rural migrants' housing as a platform for future housing development.

The main findings of the study suggest that the rural migrants' living condition were highly affected by their housing characteristics because of the existence of internal and external constraints among rural migrants. These constraints had impeded the rural migrants from pursuing good housing characteristics, thus decreasing the standard of their living condition. In order for the rural migrants to increase their housing characteristics and living condition, these constraints need to be removed or restricted.

The constraints have been identified as follows.

7.2.1 Access to Low Income Occupation

Most of the rural migrants migrate to urban area without any skills, thus their financial crises force them to do some small businesses or low income occupations. Thus, the low income causes poor housing characteristics whether in the form of rented housing or having to live in some other places. Accessibility to middle or higher income occupation would improve the rural migrants' housing characteristics, thus making their living condition improve as well.

7.2.2 Rural Migrants' Attitudes

Rural migrants migrate to urban areas, wishing to earn more money and finally being able to send the money back to their loved hometowns. Based on this, very few rural migrants have enjoyed spending money to improve their housing characteristics and living condition. Their indifferent attitude towards housing for example, housing in urban areas suggests that they see their housing only as a temporary living place. They do not care about the poor housing characteristics and living condition, as long as they have a place to sleep is already a fortunate thing. Their attitude has restricted them to improve their housing characteristics and living condition.

7.2.3 Government Inefficiency

Government intervention in rural migrants is essential. While the Beijing government has never done any supportive actions towards these people, it has been inefficient in implementing its policy to benefit them. The roles of the Beijing government as well as the department need to be revised too, so as to improve their efficiency in benefiting the rural migrants. It is however unclear whether the government's supportive policies are produced mainly for political enhancement.

Despite the constraints that have shackled the rural migrants, their slow income has some how done much convenience to local residents. Many rural migrants perform hard labour to contribute to the construction of social infrastructure, also some rural migrants do some businesses in Beijing like selling fruits, vegetables or become hawkers which further give local residents more convenience.

However, there have been many migrants (including urban and rural migrants) living and working in Beijing, thus it results in the overcrowding of public transportation (subway and bus) and sanitation problem. Every day, especially in the morning and evening, the public transportation is very crowded and this disturbs the local residents' lives. Also the businesses that they do in the streets have contributed to many sanitation problems.

Meanwhile, many rural migrants bring their children to Beijing to do business with them together. It is because they do not have the Beijing *hukou* that their children cannot attend Beijing public school, and at the same time, the tuition fees of private schools are very high. Thus, the majority of rural migrants decide to allow their children to quit school. Thus their children's future development will definitely be affected.

Until now, the Chinese government already realizes the situation of the migrants in the cities, but due to the population increase too fast together with many city problems (traffic jam, public transportation crowding, job competition, housing shortage etc.), they just want to maintain the population in the cities and want these migrants back to their hometown finally. Therefore, the Chinese government strictly controls the *hukou* (local ID) and the migrants are quite difficult to get it. The migrants only could get some low level jobs in the city that provided by the local government. Fortunately, the Chinese government move many factories to the rural area and produce many job opportunities for the rural migrants in their hometown (Bian, 2005). Thus, this action could encourage the migrants back to their hometown finally. But in the urban area, the local government must pay attention to the migrants in future and give more benefits to them like solving the housing problem, children could attend school, give more job opportunity and give them *hukou* (local ID) etc. Thus, the migrants' housing characteristics and living condition could be improved a lot.

Overall, the rural migrants have made some positive and negative contributions to Beijing. Their personal and work affairs should be cared for by the government and the whole society, and at the same time, the rural migrants' attitude should be altered to a certain degree.

7.3 The Future

Unlike the migration of the developed countries, the urban and rural migration in China carries some special circumstances. Migration is stimulated by economic functions and migrants' own willingness. The willingness compels them to migrate to big cities like Beijing, Shanghai, Shenzhen etc. to get good income and enjoy the good facilities commonly provided in big cities. Due to the large population movement in China, the local government uses *hukou* to control the migrants and not allow them to enjoy the social welfare that the local residents can enjoy. It has been seen that the role of the *hukou* could separate them from the local residents and even send them back to their hometown. While *hukou* policies can deteriorate the relationship between local and migrants, at the same time, the migrants have been unsatisfied with the local government's policy. Thus, it is inevitable that some social problems emerge in big cities.

From the economic perspective, the migrants travel to big cities to look for good occupations. Therefore, the job competition between local and migrants becomes higher. Thus many local residents have a sense of dislike towards the migrants and the relationship between them sours. Also, some social resources like bus, subway etc. tend to become crowded. Therefore, the local government and company should do some negotiations between the locals and the outsiders. For the public transportation, the local government should increase the number of buses and avoid overcrowding the best way possible.

From the policy perspective, Beijing government should care more about the migrants and give them more benefits like social welfare, occupation etc. For the *hukou* problem, the Beijing government should revise the requirement, make the policy more lenient, and help migrants to get Beijing *hukou* easily. For example, the Beijing government should endorse the Beijing *hukou* if they work in Beijing for more than one year. At the same time, the Beijing local residents should care about their welfare too and do not discriminate them.

In the future, the migrants moving to big cities like Beijing, Shanghai and Shenzhen etc. should become more and more adaptable to the Chinese situation, as the population in China increases rapidly and everyone tends to aspire living in big cities. Therefore, the governments should make more efforts on this group and some corrective measures should be undertaken to provide more assistance like *hukou*, housing, occupation, education, social welfare etc.

7.4 Limitations and Scope for Further Research

This study has been an attempt to examine the influence of housing characteristics (overcrowding, privacy and facility) on rural migrants' living condition in Beijing Fengtai District. The rural migrants' living condition is highly affected by their housing characteristics. At the same time, whether they have Beijing *hukou* or not is very important to the social welfare in Beijing. The concept is based on many writers in similar studies in different regions of the world. In order to consolidate the findings of similar studies in the future, it is suggested that other areas like Shanghai, Shenzhen, Guangzhou, Hong Kong and Singapore etc should be examined in more detail and in different housing perspectives like air ventilation, sewerage system and housing planning etc. It is hoped that a more comprehensive study of rural migrants' housing study might be developed in order to evaluate its effects on their living condition.

Limitations of cost and time prompted the researcher to choose Beijing Fengtai District as a study area. However, some aspects of the findings and recommendations proposed above should be the starting point for further research in other study areas. One of the areas of the key interests is the aspect of housing congestion of the rural migrants. In countries where housing problem is very common like Singapore and Hong Kong, interesting findings related to this have been generated. Since some researchers have already done a lot of research about housing overcrowding in these areas, it would be beneficial to all interested government officers to study and improve the housing development in China.

Another area of research which could be useful as an extension of this study is a genome-wide association study (GWAS) to the migrants' housing characteristics and living condition. A study of this type is appropriate to examine the relationship in a more comprehensive manner since the nature of such relationship is complex and dynamic.

Finally, further studies could be narrowed down to other sub-topics within the scope of migrant housing issues. Issues related to housing characteristics and living condition for example, could be developed to produce a deeper understanding of the potential of migration. A more detailed study on migrants' housing (including air ventilation, sewerage system or housing planning) would also be useful as a means of assisting the government to maximize their assistance for this group of people.

It is hoped that the findings of this study will be able to provide some guidance for improvements in the way we perceive migrant housing. The findings are also hoped to serve as a starting platform for further research, and for it to be able to shed some light on the still neglected areas of migrant housing.

References

[1] Abera, K. & Yemane, B. (2002). Crowding in a traditional rural housing ("Tukul") in Ethiopia. *Ethiopia. Journal Health Development*, Vol. 6 (3), pp. 303-308.

[2] Alwash, R. and McCarthy, M. (1988). Accidents in the home among children under 5: ethnic differences or social disadvantage. *British Medical Journal*, Vol. 296 21 May, pp. 1450-1453.

[3] Altman, I. (1975). *The Effects of Crowding and Social Behaviour*, Brooks/Cole Publishing Co., California.

[4] Altman, R. (1996). Privacy: A conceptual analysis. *Environment and Behavior* 8, pp. 7-29.

[5] Ambrose, P. (1996). *The Real Cost of Poor Homes: A Critical Review of the Literature*, University of Sussex and University of Westminster.

[6] An, L. (2006). The health condition of floating population in Beijing and Qingdao. *China Scientific Press*. Vol. 11, pp. 16-22.

[7] Asami, Y. (2005). Living condition—theory and evaluation methods. *Qinghua University Press*, Vol. 2, pp. 27-39.

[8] Babbie, E. R. (1983). The practice of social research, third edition. Belmont, *California*: Wadsworth.

[9] Baldassare, A. (1999). Residential crowding in urban America. *University of Caliifornia Press*, Berkeley.

[10] Barber, B. (2007). Social stratification. *New York: Harcourt, Brace and World*.

[11] Baron, R. M. (1996). Effects of social density in University residential environments. *Journal of Personality and Social Psychology*, Vol. 36.

[12] Baum, A. and Davis, G. E. (2006). Spatial and social aspects of crowding perception. *Environment and Behavior* 8, pp. 527-544.

[13] Baum, A. & Calesnick, L. E. (1998). Crowding and personal control: social density and the development of learned helplessness. *Journal of Personality and Social Psychology*, Vol. 7.

[14] Bateson, G. (1998). Naven: The culture of the Iatmul people of New Guinea as revealed through a study of the Naven Ceremonial. *Wildwood House*, London.

[15] Blake, K. S. & Kellerson, R. L. & Simic, A. (2007). Measuring overcrowding in housing. *U.S. Department of Housing and Urban Development, Officer of Policy Development and Research*. Vol. 1.

[16] BCD (2005). Beijing Construction Department. Beijing construction department report: Beijing housing investigation in 2005.

[17] BCD (2006). Beijing Construction Department. Beijing Construction Department report: Rural migrant housing facility in Beijing conducted in 2006.

[18] BCD (2010). Beijing Construction Department. Beijing local government report: Housing construc-

tion cannot keep up with population increase in 2010.
[19] BCD (2012). Beijing Construction Department. Beijing local government report: Population census in Beijing in 2012.
[20] BMRS (2010). Beijing Migrant Registration System. Beijing local government report: Migrant improves this year in Fengtai in 2010.
[21] BFPCO (1997). Beijing Foreign Population Census Office. Beijing foreign population census report: Foreign population census in 1997.
[22] BG (2006). Beijing Government. Beijing government report: Beijing population survey in 2006
[23] BG (2012). Beijing Government. Beijing government report: Housing policy in Beijing, 2012.
[24] BG (2012). Beijing Governmen. Beijing government report: Housing census in Beijing, 2012.
[25] BHM (2009). Beijing Housing Market. Beijing housing market: Beijing housing investigation in 2009.
[26] BHM (2012). Beijing Housing Market. Beijing local government report: High housing price in Beijing in 2012.
[27] Bharucha, R. R. and Kiyak, H. (1992). Environmental effects on affect: density, noise and personality. *Population and Environment* 5, pp. 60-72.
[28] Bian, Y. J. (2005). Social stratification, housing ownership and living condition in China. *Social and Scientific Press*, Vol. 3, pp. 23-26.
[29] Bian, Y (1994). Work and inequality in urban China. State University of New York Press, Albany. *American Sociological Review* 62, pp. 366-385.
[30] Blau, P. M. and Ruan, O. (1990). Inequality of opportunity in urban China and America. *Research in stratification and mobility* 8, JAI Press.
[31] BMRS (2010). Beijing Migrant Registration System. Beijing local government report: Migrant improves in 2010.
[32] BMBS (2006). Beijing Municipal Bureau of Statistics. Beijing local government report: Population census in Beijing in 2005.
[33] BMBS (2011). Beijing Municipal Bureau of Statistics. Beijing local government report: The average living space of Beijing residents in 2011.
[34] BMBS (2012). Beijing Municipal Bureau of Statistics. Beijing local government report: Migrant population number in recent years in Beijing.
[35] BMBS (2013). Beijing Municipal Bureau of Statistics. Beijing local government report: The average income of Beijing residents in 2013.
[36] BMCS (2012). Beijing Migration Control System. Beijing local government report: Migrants' population number increase in Beijing in 2012.
[37] Bo, G. Y. (2010). Housing policy renovation in China. *Beijing Real Estate Press*, Vol. 1, pp. 11-19.
[38] Boyd, M. (1989). Family and Personal Networks in International Migration: Recent Development and New Angendas. *International Migration Review*, 1989, 23(3).
[39] BTG (2011). Beijing Travel Guide. Beijing Travel Guide report: Beijing real estate, 2011.

[40] Catherine, D. (2007). A practical guide to research methods. *Published by How To Content*, Vol. 3.

[41] Cai, T. (2001). Intelligent housing in developed countries. *Journal of Housing* (China), Vol. 1, pp. 67-75.

[42] Cai, Y. W. (2002). Rural migrants' education level in Shanghai. *Beijing University Press*, Vol. 1, pp. 155-176.

[43] Cagamas, B. (1997). Housing the nation: a definitive study, *Cagamas Berhad*, Kuala Lumpur.

[44] CCG (1986). China Central Government. China Central government report: National housing investigation report in China, 1986.

[45] CCG (1987). China Central Government. China Central government report: Urbanization and rural migrants in 1987.

[46] CCG (1995). China Central Government. China Central government report: Housing condition survey in China, 1995.

[47] CCG (2000). China Central Government. China Central government report: Population census in China, 2000.

[48] CCG (2006). China Central Government. China Central government report: China housing development plan in 2006.

[49] CCG (2007). China Central Government. China Central government report: Housing census in China, 2007.

[50] CCG (2009a). China Central Government. China Central government report: China housing development plan in 2009.

[51] CCG (2009b). China Central Government. China Central government report: Low-rent housing investigation in 2009.

[52] CCG (2009c). China Central Government. China Central government report: Population census in China, 2009.

[53] CCG (2010). China Central Government. China Central government report: Population census in China, 2010.

[54] Chan, M. K. (1997). Privacy: does it concern the people. *Hong Kong Institute of Asia-Pacific Studies*, pp. 359-381.

[55] Chan, K. W. and Zhang, L. (1999). The *hukou* system and rural-urban migration in China: processes and changes. *China Quarterly*, Vol. 1, pp. 18-55.

[56] Chen, B. Z. (2003). Living condition of the sustainment development in China. *Sciencific Press*. Vol. 1, pp. 69-77.

[57] Chen, D. Y. (2003). Housing development change the history. *Journal of Housing* (China), Vol. 1.

[58] Chen, Y. (2002). The comparison of housing condition in urban and rural area of China. *Xin Hua Press*, Vol. 1.

[59] Chen, Y. X. (2005). The energy conservation inside housing in US. *Journal of Housing* (China), Vol. 2.

[60] Cheng, X. H. (2011). Real estate law. *Beijing People Press*, Vol. 2.

[61] Chan, K. W. (1994). Cities with invisible walls: urbanization in post-1949 China. Oxford University Press, Hong Kong.

[62] Chan, A. M. (1996). China's urban housing reform: price-rent ratio and market equilibrium. *Urban Studies* 35, Vol. 1.

[63] Chen, X. M. and Gao, X. Y. (1993). Urban economic reform and housing investment in China. *Urban Affairs Quarterly* 29, Vol. 1.

[64] Chiu, R. (1996). Housing affordability in Shenzhen special economic zone: a forerunner of China's housing reform. *Housing Studies* 11, pp. 567-580.

[65] CMC (2010). China Ministry of Construction. China Ministry of Construction report: The record of housing construction in China, 2010.

[66] CMF (2007). China Ministry of Finance. China Ministry of Finance report: The real estate of housing investigation, 2007.

[67] Cui, J. Y. (2011). Housing real estate research in China. *China Law Press*, Vol. 1.

[68] Davis, D. (1990). Urban job mobility. *Chinese society on the eve of Tiananmen*, Harvard University Press, Cambridge, CA.

[69] Ding, J. H. (2001). The crime of rural migrants in Shanghai. *Population Research*, Vol. 6.

[70] Ding, J. H. (2005). Rural migrants housing investigation in Shanghai. Housing Investigation report.

[71] Duan, C. R. (2009). The century of floating population in Beijing. *Beijing Scientific Press*, Vol. 1, pp. 11-12.

[72] Epstein, A. L. (1994). Privacy and the boundaries of the self: reflections on some Tolai data. *Canberra Anthropology* 17(1), pp. 1-29.

[73] FI (1957). France Institute. Housing research report in France: The research of living condition in France, 1957.

[74] Feng, J. (2009). Housing and residential policy. *China Construction and Industry Press*, Vol. 1.

[75] Feng, W. P. (1999). About 2000 houses were sold to non-Beijing residents in 1998. *Beijing Press* 1, pp. 36-47.

[76] Feng, G. L. (1997). The consumer behavior of rural migrants in urban area. *Jiang Han Press*, Vol. 4, pp. 5-17.

[77] Freedman, J. L. (1971). The effect of crowding on human task performance. *Journal of Applied Social Psychology*, Vol. 1.

[78] Freedman, J. L. L. & Buchanan, R. W. P. J. (1972). Crowding and human aggressiveness. *Journal of Experimental Social Psychology*, Vol. 1.

[79] Freedman, J. L. L. (1975). Crowding and Behavior. *Freeman*, SanFrancisco.

[80] Fu, Q. L. (2009). Urban real estate development in China. *Law Press*, Vol. 1, pp. 26-29.

[81] Gao, P. Y. (1993). International housing policy and management. *Beijing People Press*, Vol. 1, pp. 10-21.

[82] Gaubatz, P. R (1995). Urban transformation in post-Mao China: impacts of the reform era on

China's urban form. *Woodrow Wilson Center Press and Cambridge University Press* 1, Cambridge.

[83] Glaeser, B. (1995). Housing, sustainable development and the rural poor: a study of Tamil Nadu, New Delhi: Sage Publications.

[84] Goldscheider, G. (1983). Urban migration developing nations. *Westview*.

[85] GM (2012). Google Map. Beijing Fengtai District location.

[86] Gu, C. L. (2002). Urban social studies. *Dongnan University Press*, Vol. 1, pp. 30-35.

[87] Guo, J. P. (2004). Social stratification in urban area. *Academic Forum*, Vol. 1, pp. 89-90.

[88] Guo, Y. C. (2006). The living space of floating population in big cities. *Gepgraphy Research and Development*, Vol. 5, pp. 76-78.

[89] Havel, J. E. (1957). Living and housing theory. *France of University Press*, Vol. 1.

[90] Habitat, A. (1991). Graphic presentation of basic human settlements statistics. *UNCHS (Habitat)* Nairobi, Kenya.

[91] Hao, Y. D. (2009). Housing development theory. *China Construction Industry Press*, Vol. 2, pp. 17-20.

[92] Harris, R. (1991). "A Working-Class Suburb for Immigrants, Toronto, 1909-1931." *Geographical Review* 81 (1991).

[93] He, D. G. (2005). The energy conservation housing in Japan. *Journal of Housing in China*, Vol. 1, pp. 126-130.

[94] He, M. X. (2008). The housing condition of rural migrants in Beijing. *China People Press*, Vol. 1, pp. 69-70.

[95] Hertzog, M. A. (2008). Considerations in determining sample size for pilot studies. *Research in Nursing and Health*, Vol. 31, pp. 180-191.

[96] Hendershot & Gerry, E. (1978). The dynamics of migration: Internal migration and migration fertility. *Interdisciplinary Communications Program*, Vol. 1.

[97] Huang, Y. Q. (2003). Renter's housing behavior in transitional urban China. *Housing Studies*, Vol. 18, pp. 103-126.

[98] Huang, J. (2010). Housing and finance system in China. *Journal of Housing* (China), Vol. 1, pp. 9-17.

[99] Hu, D. H. (2009). China real estate. *Sichuan University Press*, Vol. 1, pp. 15-23.

[100] ICF (2005). ICF International analysis of AHS data, 1985~2005. U.S. Department of Housing and Urban Development report: Measuring overcrowding in housing.

[101] Insel, P. M. & Lindgren, H. C. (1998). Too close for comfort: The psychology of crowding. Prentice-Hall, *Englewood Cliffs*, NJ.

[102] Jeanne, W. and Bonnie, D. (2001). Housing characteristics in US, 2000. Department of Commerce, Economics and Statistics Administration, *US Census Bureau*, Vol. 2, Issue 8, pp. 56.

[103] Jiang, L. W. (2006). Living condition of floating population in urban China. *Housing Studies*, Vol. 21, No. 5, pp. 719-744.

[104] Jian, F. S. & Ye, F. H. (2003). The working and living space of the "floating population" in China. *Asia Pacific Viewpoint*, Vol. 44, No. 1, April 2003.

[105] Knodel, Chayovan, J. N., and Siriboon, S. (1992). The impact of fertility decline on familial support for the elderly: An illustration from Thailand. *Population and Development Review* 18(1), pp. 79–103.

[106] Krivo, L. J. (1995). Immigrant characteristics and Hispanic-Anglo housing inequality. *Demography* 32, pp. 599–615.

[107] Lau, K. Y. (1993). Urban housing reform in China, amidst property boom year. *China review 1993*. Chinese University Press, Hong Kong.

[108] Leaf, M. (1997). Urban social impacts of China's economic reforms. *Cities* 15, Vol. 1.

[109] Leopore, et al. (1992). Role of control and social support in explaining the stress of hassles and crowding. *Environment and Behavior* 24, pp. 795–811.

[110] Li, M. B. (1991). The effects of floating to urbanization. *Economy Newspaper Press*, Vol. 1.

[111] Lim, G. C. and Lee, M. H. (1990). Housing policy in modern China. *Environment and Planning C: Government and Policy* 8, pp. 477–487.

[112] Lin, N. and Bian, Y. (1991). Getting ahead in urban China. *American Journal of Sociology* 97, pp. 657–688.

[113] Liu, C. (2004). The family investigation of floating population in urban area. *Migration Research*, Vol. 7.

[114] Lin, L. Y. (2003). Housing conditions of the floating population under the double residential status and the factors affecting them: a case study in Fujian Province. *Ministry of Education of China*, Vol. 1.

[115] Lin, J. & Feng, C. C. (1998). Housing policy in oversea. *Journal of Urban Planning*, Vol. 2.

[116] Liu, Z. R. (1995). Housing innovation in Indian. *Journal of Housing in China*, Vol. 3, pp. 18–20.

[117] Li, H. P. (2000). Housing policy in Germany. *Journal of Housing Construction* (China), Vol. 1.

[118] Lipset, S. M. and Bendix, R. (2001). Social mobility in industry society, Berkeley/Los Angeles: *University of California Press*.

[119] Li, Q. (2004). Rural migrants and social stratification. *Social and Scientific Press*, Vol. 1, pp. 22–27.

[120] Li, X. F. (2000). Rural migrants in urban area in China. *Real Estate Newspaper Press*, Vol. 10, pp. 37–56.

[121] Li, W. (2004). Compare the housing price and income in China and International. *Liaoning Economic Press*, Vol. 5, pp. 29–37.

[122] Li, W. D. (2006). The relationship between rural migrant and urbanization. *Urban Issue*, Vol. 8.

[123] Liu, X. (2005). The aspects of housing energy conservation in Germany. *Journal of Housing* (China), Vol. 1.

[124] Liu, Z. C. (2007). The living condition and environment of floating population in urban area. *Xi An University Press*, Vol. 2.

[125] Lin, P. (2005). The energy conservation housing in future. *Journal of Housing* (China), Vol. 1, pp. 79–93.

[126] Loo, C. (1973). Important issues in researching the effects of crowding on humans. *Representative Research in Social Psychology*, Vol. 3.

[127] Logan, J. R. (1993). Inequalities in access to community resources in a Chinese city. *Social Forces* 72, pp. 55-76.

[128] Logan, J. R. (1996). Market transition of housing in urban China. *International Journal of Urban and Regional Research* 20, pp. 400-421.

[129] Lu, X. Y. (2010). The population movement in China. *Social Scientific Press* (China), Vol. 1, pp. 11-12.

[130] Lu, Q. (2005). The floating population in Beijing and the effects to economy development. *Social and Scientific Press*, Vol. 5.

[131] Luo, P. (2007). The housing system in urban China. *Qinghua University Press*, Vol. 1, pp. 10-11.

[132] Luo, R. C. (2009). Social structure of floating population in Shanghai. *Tong Ji University Press*, Vol. 1.

[133] Lue, L. M. (2009). Land and Market. *Gui Zhou People Press*, Vol. 1.

[134] Massey, D. S. (1986). The Social Organization of Mexican Migration to the United States. *Annals of the American Academy of Political and Social Science*, Vol. 1.

[135] Ma, X. (1998). The migration period in China. *Beijing Economic Press*, Vol. 1, pp. 57-58.

[136] McArdle, N. & Mikelson, K. S. (1994). The new immigrants: Demographic and housing characteristics. Working Paper W94-1. Cambridge, MA: *Joint Center for Housing studies*.

[137] Mcgrew, W. C. (1970). Social and spatial density effects on spacing density in pre-school children. *Journal of Child Psychology and Psychiatry* 11, pp. 135-176.

[138] Meng, Q. J. (2009). The research of living condition of floating population in Shanghai. *Shanghai Social and Scientific Press*, Vol. 1, pp. 56-57.

[139] Michael, H. S., Samantha, F. and Emily, R. (1998). The housing conditions of immigrants in New York City. *Journal of Housing Research*, Vol. 9, Issue 2.

[140] Mitchell, D. (1996). The life of the land: Migrant Workers and the California Landscape. Minneapolis: *University of Minnesota Press*.

[141] Myers, D., Baer, W. C. and Choi, S. Y. (1996). The changing problem of overcrowded housing. *Journal of the American Planning Association* 62, pp. 66-84.

[142] Nattrass, N. (1997). Street trading in Transkei-A struggle against poverty, persecution and prosecution. *World Development*, Vol. 15, No 7, pp. 861-875.

[143] Nee, V. (1996). The emergence of a market society: changing mechanisms of stratification in China. *American Journal of Sociology* 101.

[144] Pang, L. H. (2003). Migration and housing condition in China. Paper presented at *Conference on Scientific Study of China 5th Census*, March, pp. 28-31, 2003, Beijing.

[145] Pader, E. (1994). Spatial relations and housing policy: Regulations that discriminate against Mexican-Origin households. *Journal of Planning Education and Research* 13 (2), pp. 119-135.

[146] PHCE (1984). Population and Housing Census of Ethiopia. Analytical report at national level,

Addis Ababa, pp. 340-353.
[147] Portes, A. (1995). Economic sociology and the sociology of immigration: a conceptual overview. New York, *Russell Sage Foundation*.
[148] Qian, Q. L. (2003). The migration in Beijing. *Urban Planning*, Vol. 11.
[149] Qian, X. F. (2003). The investigation of rural migrants in Nanjing. *Nan Jing Social Scientific Press*, Vol. 9.
[150] Rapoport, A. (1976). Toward a redefinition of density. In Crowding in Real Environments, ed. S. Saegert, pp. 7-26. *Sage, Beverly Hills, CA*.
[151] Robert, V. K. & Daryle, W. M. (1970). Determining sample size for research activities. *Educational and Psychological Measurement*, Vol. 30, pp. 607-610.
[152] Rodin, J. (1976). Density, perceived choice and responses to controllable and uncontrollable outcomes. *Journal of Experimental Social Psychology*, Vol. 12.
[153] RUC (2006). Renmin University of China. Renmin University Report: Population research in China, 2006.
[154] Sababu, K. (1998). Privacy and crowding concepts in Melanesia: the case of Papua New Guinea. *Habitat International*, Vol. 22, No. 3, pp. 281-298.
[155] Sanders, P. (1986). Social theory and the urban question. Vol. 2, London.
[156] SCD (2006). Shanghai Construction Department. Shanghai Construction Department report: Rural migrant housing facility survey in Shanghai conducted in 2006.
[157] Shen, J. Z. (2006). Housing policy in Japan and the implication to China. *Housing Studies* (China), Vol. 1.
[158] Sherrod, D. R. (1974). Crowding, perceived control and behavioral affect effects. Journal of *Applied Social Psychology*, Vol. 4.
[159] Smith, M. J. (1991). Crowding, task performance and communicative interaction in youth and old age. *Human Communication Research* 7, pp. 259-272.
[160]
[161] Stanley, D. C. (1971). The history of working-class housing: a symposium. *Newton Abbot: A Symposium*, pp. 168-225.
[162] Stephen, W. K. Mak, Lennon, H. T. Choy and Winky, K.O. Ho. (2007). Privatization, housing conditions and affordability in the People's Republic of China. *Habitat International*, Vol. 31, Issue 2, pp. 177-192.
[163] Stokols, D. (2002). On the distinction between density and crowding: some implications for future research. *Psych. Review*, pp. 79-91.
[164] Tan, L. (2003). The gender analysis of informal occupation of floating population in China. *Population Research*, Vol. 5, pp. 15-38.
[165] Tian, D. H. (1998). Compare the International housing policy and China. *Qinghua University Press*, Vol. 1, pp. 79-95.
[166] Todaro, M. P. (1969). A model of labor migration and urban unemployment in LDCs. *American Economic Review*, Vol. 1.

[167] Tolley, G. (1991). Urban housing reform in China: an economic analysis. World Bank Discussion Paper 123, The World Bank, Washington DC.

[168] UNHSP (1976). The United Nations Human Settlements Program. United Nation report: International Living Condition Standard, 1976.

[169] UN (2007). United Nations. United Nations report: World Urbanization Prospects, 2007.

[170] USDHUD (2007). U.S. Department of Housing and Urban Development. U.S. Department of Housing and Urban Development report: Measuring overcrowding in housing, 2007.

[171] Valins, S. (2001). Residential group size, social interaction and crowding. *Environment and Behavior*, Vol. 5.

[172] Vogel, E. (1990). One step ahead: Guangdong under reform. Harvard University Press, Cambridge, MA.

[173] Walden, T. A. (1981). Crowding, privacy and coping. *Environment and Beh*avior 13, pp. 205–224.

[174] Wang, Y. (1990). Private sector housing in urban China since 1949: the case of Xi'an. *Housing Studies* 7.2, pp. 119–137.

[175] Wang, Y. P. (1992). The development and planning of Xian since 1949. *Planning Perspectives* 7, pp. 1–26.

[176] Wang, Y. P. (1996). The process of commercialization of urban housing in China. *Urban Studies* 33, pp. 971–989.

[177] Wang, C. G. (1995). Social movement and re-structure. *Zhe Jiang People* Press, Vol. 1.

[178] Wang, C. L. (2007). The analysis of floating population's movement in Shanghai. *Nan Jing Management*, Vol. 4, pp. 37–59.

[179] Wang, D. (2006). Living condition of migrants in big cities in China. *Beijing University Press*, Vol. 1.

[180] Wang, G. X. (2005). The living status of floating population in Shanghai. *Shanghai Meeting*.

[181] Wan, M. (2005). The consuming behavior and strategy selection of floating population in Beijing. *Beijing Commercial University Press*, Vol. 3, pp. 11–29.

[182] Wang, Q. (2007). The life style of rural migrants in Guangzhou. *Social and Scientific Press*, Vol. 2, pp. 75–93.

[183] Wang, R. (2007). Study of living condition of floating population in China. *The Journal of Zhongnan University of Economic and law*, Vol. 2.

[184] Wang, Y. P. (2000). Social and spatial implications of housing reform in China. *International Journal of Urban Regional Research* 24, pp. 1–19.

[185] Wang, W. (1999). Housing and living facility in urban cities. *Beijing Renmin People University Press*, Vol. 1.

[186] Wang, X. Y. (2003). Housing improvement in last ten years. *China Real Estate*, Vol. 1.

[187] Wang, Z. H. & Huang, K. K. (1999). Rural migration and urbanization in Europe and US. *Beijing Social Science Press*, pp. 199–256.

[188] Wang, W. G. & Wang, G. H. (2002). Land and housing construction in China. *China Zheng Fa*

University Press, Vol. 1.
[189] Wang, J. F. & Huang, M. C. (2011). Land law theory and practice. *People Daily Press*, Vol. 1.
[190] Wang, X. Z. (2004). Living structure in urban area in China. *Scientific Press*, Vol. 1.
[191] Wang, Y. (2014). Housing system in the world. *Scientific Press*, Vol. 1.
[192] Westin, A. (1970). Privacy and freedom. New York: *Atheneum*.
[193] World Bank (1992). *China: implementation options for urban housing reform*. A World Bank country study. The World Bank, Washington DC.
[194] Wu, F. L. (1996). Changes in the structure of public housing provision in urban China. *Urban Studies* 33, pp. 16-27.
[195] Wu, L. F. & Luo, D. L. (2009). *Housing policy in China*. Vol. 1, pp. 206-210.
[196] Wu, X. (2010). Analysis the living space of floating population in big cities of China. *Dong Nan University Press*, Vol. 1.
[197] Wu, X. & Zhang, J. (2002). The public housing in Hong Kong and Singapore. *Journal of Urban Planning*, Vol. 3.
[198] Wu, W. P. & Wang, H. S. (2002). The housing condition analysis of floating population in Beijing and Shanghai. *Beijing University Press*, Vol. 1, pp. 2.
[199] Wu, Z. T. (1996). Housing and living environment in Germany. *Housing and Real Estate*, Vol. 1.
[200] Xiang, Y. (2003). The energy conservation in UK. *Journal of Housing* (China), Vol. 1, pp. 16-18.
[201] Xiao, C. (1993). The consideration to the real estate in China. *Urban city Press*, Vol. 1, pp. 5-9.
[202] Xie, Y. (1996). Regional variation in earnings inequality in reform-era urban China. *American Journal of Sociology* 101, pp. 950-992.
[203] Xie, J. J. (2009). Real estate development in China. *China Market Press*, Vol. 1.
[204] Xu, Q. (1993). Compare the International housing policy. *Housing and Real Estate*, Vol. 5.
[205] Xu, W. (2005). Population migration research. *Shanghai Migration Research Press*, Vol. 6.
[206] Yang, L. and Wang (1992). Housing reform: theoretical rethinking and practical choices. People's Press of Tianjin, Tianjin.
[207] Yang, F. & Yan, X. P. (2000). To analyze the living condition in urban area in China. *Urban Issue*, Vol. 4.
[208] Yang, J. R. (2000). Thirty years economic development in China. *Xi Nan University Press*, Vol. 1.
[209] Yang, J. R. (2012). The record of welfare housing system in the elementary stage of China. *Journal of Housing* (China), Vol. 1.
[210] Yang, L. & Wang, Y. K. (2006). Housing renovation in China. *Tianjin People Press*, Vol. 1.
[211] Ye, B. (1993). Residential housing in Shanghai: 1949-1990. Scientific Press of Shanghai, Shanghai.
[212] Yi, Z. G. & Hong, X. L. (2008). The movement tendency of the floating population in Beijing.

Beijing Press, Vol. 6.

[213] Ying, K. C. (1998). Density, crowding and factors intervening in their relationship: evidence from a hyper-dense metropolis. *Social Indicators Research* 48, pp. 103-124.

[214] Yi, D. (1997). Public housing in Hong Kong. *Journal of International Housing*, Vol. 6, pp. 16-30.

[215] Yu, M. (2003). The research of housing status in developed countries. *Journal of Housing* (China), Vol.1.

[216] Zhan, Y. (2003). Living condition and income between migrants. *Urban Issue*, Vol. 4.

[217] Zhao, R. & Yue, S. Z. (2005). The research on floating population of the social and security guarantee. *Development Research*, Vol. 6.

[218] Zhang, X. Q. (1997). Chinese housing policy 1949-1978: the development of a welfare system. *Planning Perspective* 12, pp. 433-455.

[219] Zhang, W. L. (1997). Social benefits of rural migrants in cities. *Zhe Jiang Academic Press*, Vol. 2.

[220] Zhang, Z. X. & Hou, Y. F. (2009). Floating population in urban communities. *Social Science Academic Press* (China), Vol. 1.

[221] Zhai, Z. W. & Zhang, Y. (2003). The analysis of housing condition of floating population in China. *Conference Paper of China fifth population census*.

[222] Zhang, E. Z. (2010). Housing and finance renovation in China. *Housing and Finance Research in China*, Vol. 1.

[223] Zhang, Q. J. & Du, D. B. (2000). Housing and population issue in US. *Urban Research*, Vol. 3.

[224] Zhang, Y. M. (2007). The living condition of urban migrants in Lanzhou. *Social and Scientific Press*, Vol. 1.

[225] Zhang, X. H. (1995). Urban migration management and strategy. *Journal of Strategy and Management*, Vol. 6.

[226] Zhao, C. C. (2005). The housing energy conservation in Finland. *Journal of Housing* (China), Vol. 1.

[227] Zheng, Q. M. (2004). The peasants' living factor in urban area. *Social Geography Press*, Vol. 1.

[228] Zhong, S. Y. (2000). Population movement and social economy development. *Wuhan University Press*, Vol. 1.

[229] Zhou, W. W. (2002). Migrant family in China. *Hebei People Press*, Vol. 1.

[230] Zhou, Y. Q. (2008). Residential quality of Chinese urban households. *Social Science Academic Press* (China), Vol. 1, pp. 89-97.

[231] Zhou, L. C. (1996). The floating population in Beijing. *China Population Press*, Vol. 1, pp. 3-11.

[232] Zhou, C. C. (2009). Housing quality and measurement. *Beijing Social Science Press*.

[233] Zhu, X. Q. (1993). Shanghai's housing 1949-1990. *Shanghai Science Press*, Shanghai.

[234] Zhu, C. G. (2001). The urban migrants' characteristics in China. *Population Research Press*, Vol. 2.

[235] Zhou, D. M. (2003). Rural peasants in urban area. *China Population Press*, Vol. 1, pp. 19-30.
[236] Zhou, M. W. (2000). International living quality assessment. *Housing in Shanghai*, Vol. 7.
[237] Zhou, X. H. (1998). The effects of floating population to urban cities in Beijing and Wenzhou. *Social Research*, Vol. 5, pp. 67-76.
[238] Zhu, Y. (2007). China's floating population and their settlement intention in the cities: Beyond the *Hukou* Reform. *Journal of Housing Research*, Vol. 31 (1), pp. 65-76.
[239] Zhu, L. (2002). The adaption of rural migrants in urban area. *Jianghai Press*, Vol. 6.
[240] Zong, C. F. (2007). The investigation of rural labor in Nanchang. *China Rural Investigation*, Vol. 1.